TURNER ON BIRDS

TURNER ON BIRDS:

A SHORT AND SUCCINCT HISTORY

OF THE

PRINCIPAL BIRDS NOTICED BY PLINY AND ARISTOTLE

FIRST PUBLISHED BY

DOCTOR WILLIAM TURNER,

1544.

EDITED,

WITH INTRODUCTION, TRANSLATION, NOTES, AND APPENDIX,

BY

A. H. EVANS, M.A.

CLARE COLLEGE, CAMBRIDGE.

CAMBRIDGE:

AT THE UNIVERSITY PRESS

1903

CAMBRIDGE
UNIVERSITY PRESS

University Printing House, Cambridge CB2 8BS, United Kingdom

Published in the United States of America by Cambridge University Press, New York

Cambridge University Press is part of the University of Cambridge.

It furthers the University's mission by disseminating knowledge in the pursuit of education, learning and research at the highest international levels of excellence.

www.cambridge.org
Information on this title: www.cambridge.org/9781107663824

First published 1903
First paperback edition 2014

A catalogue record for this publication is available from the British Library

ISBN 978-1-107-66382-4 Paperback

PREFACE.

IN translating this treatise it has been thought advisable to adhere as closely as possible to the original text of Turner, though in many places a more modern style of phraseology would doubtless have better pleased the ear of the non-scientific reader.

Not a few difficult points of interpretation have arisen in the course of the work, and the Editor welcomes this opportunity of expressing his gratitude to Professor Newton and to Mr R. D. Archer-Hind of Trinity College for the invaluable help that he has received from them in elucidating the hard passages and in revising the proofs. The Editor's thanks are also due to the Syndics of the University Press for undertaking the present publication.

9, HARVEY ROAD,
CAMBRIDGE.
July, 1903.

NOTE.

It may be remarked that the pages of Turner's work are not numbered in the original; but, for convenience of reference, the pagination is marked in the margin—the numbers being included in square brackets.

INTRODUCTION.

WILLIAM TURNER, author of the rare treatise here republished, was a native of Morpeth in Northumberland and is supposed to have been the son of a tanner of that town. By the aid of Thomas, the first Lord Wentworth, he was enabled to enter Pembroke Hall in the University of Cambridge, where he graduated B.A. and was elected a fellow of his College in 1530. At Pembroke he became acquainted with Ridley (who instructed him in Greek) and Latimer, two of the most earnest advocates of the Reformed doctrines, which he himself, both then and afterwards, strenuously embraced; but there is no need to dwell upon his theological views or the polemical works in which they were set forth.

While at Cambridge Turner was a zealous student of botany, and in 1538 published a *Libellus de re herbaria.* About two years later he left this University for Oxford, and soon after suffered imprisonment for preaching without a licence. On his release he quitted England, and travelled by way of the Netherlands and Germany to Italy, attending the botanical lectures of Luca Ghini at Bologna, where, or at Ferrara, he took the degree of M.D.

Thereafter he proceeded to Switzerland, forming a close friendship with the great naturalist Conrad Gesner of Zurich, Professor of Medicine and Philosophy in the School of that city, who held him in high esteem, and with whom he afterwards kept up a correspondence. He seems to have been at Basel in 1543, but early in 1544 he was at Cullen (Cologne), where he published not only the present work—dedicated

to Edward Prince of Wales (afterwards King Edward the Sixth)—but also edited the *Dialogus de Avibus* of his friend Gybertus Longolius of Utrecht, who died the preceding year. He pursued his botanical studies in several parts of Germany, as well as in the Netherlands, including East Friesland, for he became physician to the Count of Emden, and visited the islands of Juist and Norderney lying off the coast of that province.

On the death of King Henry the Eighth he returned to England, becoming chaplain and physician to Lord Protector the Duke of Somerset; but he lived at Kew, where he established a botanic garden. He was, moreover, incorporated M.D. of Oxford, and was appointed a prebendary of York. In 1550 the Privy Council sent letters for his election as Provost of Oriel College in Oxford, but the post had been already filled, and a similar disappointment awaited him in regard to the Presidency of Magdalen College in the same University. He then applied to Sir William Cecil for leave to return to Germany, but was soon after consoled by being appointed Dean of Wells, and, having in 1551 published the first part of his *New Herbal*, was during the next year ordained priest by his old friend Ridley, then Bishop of London.

On the accession of Queen Mary Turner had to vacate his deanery, and betook himself for safety once more to the Continent, visiting Rome and several places in Germany and Switzerland. When Queen Elizabeth ascended the throne he returned to England, and recovered his deanery, to which was attached the rectory of Wedmore in Somersetshire; but in 1564 he was suspended for nonconformity and seems to have come to live in London. In 1557 he had addressed a letter on English Fishes to Gesner, which was included in that naturalist's *Historia Animalium*; and in 1562 he published the second part of his *Herbal*, which he dedicated to Lord Wentworth, the son of his original benefactor. On the 7th of July 1568 Turner died at his house in Crutched Friars in the City of London, and was buried in the church of St Olave, Hart Street.

Turner married Jane, daughter of George Ander, Alderman of Cambridge, by whom he had issue Winifred, Peter and Elizabeth.

It must be understood that, his scientific work apart, nearly the whole of Turner's life was spent in religious controversy, and he published a considerable number of polemical works, the titles of which may be seen in the bibliography appended to the excellent 'Life' prefixed to Mr Jackson's facsimile reprint of the *Libellus de re herbaria*[1], whence all the particulars above given are taken. Other lists of Turner's works may be found in Cooper's *Athenae Cantabrigienses* (I. pp. 257—259) and the *Dictionary of National Biography* (LVII. pp. 365, 366).

Turner's object in writing the present treatise is fully set forth in his 'Epistola Nuncupatoria' prefixed to it. While attempting to determine the principal kinds of birds named by Aristotle and Pliny, he has added notes from his own experience on some species which had come under his observation, and in so doing he has produced the first book on Birds which treats them in anything like a modern scientific spirit and not from the medical point of view adopted by nearly all his predecessors; nor is it too much to say that almost every page bears witness to a personal knowledge of the subject, which would be distinctly creditable even to a modern ornithologist.

This knowledge is especially evident in his account of the habits of the Hobby (p. 19), Hen-Harrier (p. 19), Water-Ousel (p. 23), Moor-Buzzard (p. 33), Osprey (p. 37), Godwit (p. 45), Wheatear (p. 53), Sandpiper (p. 57), Fieldfare (p. 59), Cuckoo (p. 69), Black-headed Gull (p. 77), Black Tern (p. 79), Swallows (p. 101), Cormorant (p. 111), Shrike (p. 119), Redbreast and Redstart (p. 157); while his keen eye for distinctions is shown in his descriptions of the Black Cock and Grey Hen (p. 43), Godwit (p. 45), Tree-Creeper (p. 53),

[1] *Libellus de re herbaria novus*, by William Turner, originally published in 1538, reprinted in facsimile, with notes, modern names, and a Life of the Author, by Benjamin Daydon Jackson, F.L.S. *Privately Printed.* London: 1877.

Doves (p. 59), Lapwing (p. 77), Nutcracker (p. 95), Reed-Bunting (p. 103), Kites (p. 117), Bullfinch (p. 161) and others. He is most careful to tell us whether he observed the various species in England or abroad and their comparative abundance, and to note the breeding of the rarer species, such as the Spoonbill (p. 151), and Crane (p. 97) within our islands—that of the Crane being of special interest; the whole account of the Cuckoo (p. 69) is also most noticeable, as is that of the curious Walt-rapp (p. 95) of which Gesner writes as follows:

GESNER *De Corvo Sylvatico*[1] (p. 337).

AUIS, cuius hîc effigies habetur, à nostris nominatur uulgo ein *Waldrapp*; id est coruus sylvaticus...Sunt qui phalacrocoracem hanc auem interpretentur, quoniam & magnitudine & colore ferè coruum refert: & caluescit, ut uidi, cum adultior est. Turnerus Aristotelis coruum aquaticum & Plinij phalacrocoracem, & coruum syluaticum nostrum auem unam esse arbitratur, tertium genus graculi. Coruus syluaticus Heluetiorum, inquit, auis est corpore longo et ciconia paulò minore, cruribus breuibus, sed crassis: rostro rutilo, parum adunco (curuo) & sex pollices longo: alba in capite macula, & ea nuda, si bene memini....Sic ille.

Doubtless Turner's work is not free from errors, as in the case of the very old story of the breeding of the Bernicle-Goose (which, however, he was most loth to credit even when assured of its truth by an Irish Divine), in his confounding of the *Onocrotalus* with the *Ardea stellaris* and the Cornish with the Alpine Chough; yet these are but small blots on a very excellent treatise, which compares most favourably with other writings of his time.

It is quite evident from various passages that Turner was acquainted with Aristotle's works in the original Greek, and especially with his *History of Animals*; but he preferred quoting that author from the Latin translation of Theodorus

[1] Conradi Gesneri Tigurini medici & Philosophiæ professoris in Schola Tigurina Historiæ Animalium Liber III. qui est de Auium natura. Tiguri apud Christoph. Froschovervm, Anno M.D.LV.

Gaza of Thessalonica, the most celebrated Scholar of his day, who, fleeing from the sack of Constantinople, played a conspicuous part in the rise of the "New Learning," and after a course of teaching in Rome, entered successively the service of the Popes Nicholas the Fifth and Sixtus the Fourth, eventually dying in poverty in Lucania about 1484.

Exact transcription of a text was considered by no means necessary in those days: consequently we find many observations and explanations inserted in the text of Aristotle and Pliny, which had no place in the original[1].

Besides referring to Gesner, Turner mentions other learned men by name and occasionally quotes from their works; while his pages also inform us of many places that he visited.

The following excerpts from Gesner not only give instances of correspondence between him and Turner, but also shew that the former was accustomed to correct the latter from his wider knowledge of Ornithology.

De Branta vel Bernicla... (p. 107).

Idẽ [Turnerus] post librum suum de avibus publicatum, in epistola ad me data, Berniclas siue Brantas (inquit) ex putridis nauis malis fungorum more nasci, minimè fabulosum esse doctorum & honestorum uirorum oculata fides mihi persuasit. Brantà anserem palustrem ualde refert: his tamen notis ab eo differt. Branta breuior est, à collo quod rubescit nonnihil, ad medium usq; uentrem, qui candicat, nigra est. anserum more segetes populatur. In Vuallia (quæ pars est Angliæ) in Hibernia & Scotia aues istæ adhuc rudes & implumes in littore, sed non sine forma certa & propria auis passim inueniuntur. Et rursus, Præter brantam aut berniclam est alia auis, quæ originem suam arbori refert acceptam. Arbores sunt in Scotia ad littus maris crescentes, è quibus prodeunt ueluti fungi parui, primum informes, postea paulatim integram auis formam acquirunt, perfectæ tandem magnitudinis illæ, rostro aliquantisper pendent, paulò post in aquam decidunt, & tum demum uiuunt. Hoc tot tantæq; integritatis uiri affirmauerunt ut credere audeam, & aliis credere suadeã. Hæc ille. Eliota Anglus &c....

[1] The precise references to Aristotle and Pliny are now supplied, from the texts of Aubert and Wimmer, and Sillig respectively.

De Vulpansere (p. 156).

Chenalopex (inquit Turnerus Anglus) ab ansere & uulpe nomen habet...

Et rursus in epistola ad me, Vulpanserem Angli vocāt a Bergander, nidulatur in cuniculorū foueis more uulpium, anate maior, minor ansere, alis ruffis. Eliota Anglus...

De Pygargo (p. 199).

Pygargum (inquit Turnerus) literatores quidam ineptè trappum à Germanis dictum (tardam, uel bistardam) interpretantur. Sed pygargus Anglorum lingua, nisi fallar, erna vocatur, an erne. Ego ernam audio dici genus aquilę quod apud Frisios ad Oceanum Germanicum per hyemem degat, colore nigro, quod cornices quædam ut ex escarum eius reliquijs uictitent sequantur. Pygargus est forte quam Anglicè dicimus ringetayle, Eliota. Sed Turnerus ringtalum Anglis dictum ab albo circulo caudam circumeunte, buteone minorem, subbuteonem Aristotelis esse suspicatur. Quod si minor est buteone, non poterit esse pygargus.

De Caprimulgo (p. 235).

[Having quoted Turner] Idem postea in litteris ad me missis, caprimulgum se uidisse scribit prope Bonnam (Germaniæ ciuitatem ad ripam Rheni, supra Coloniam) ubi à uulgo appellatur *Naghtrauen*, id est coruus nocturnus. Nos auis illius quæ Argētinę vocatur *Nachtram*, corrupto forsan nomine, alibi *Nachtrap*, effigiem infrà ponemus cum historia nycticoracis.

De Carduele (p. 235).

...Auis Aristoteli thraupis, θραυπὶς, dicta à Gaza carduelis conuertitur: quod & Hermolao probatur. Turnero quidem non assenserim, qui chloridem nostram (*Grünling* uulgò vocant) thraupin esse conijcit.

De Rala terrestri (pp. 481—482).

...Perdix rustica vel rusticula Plinij (inquit Turnerus in epistola ad me) ab Anglis vocatur rala. Est autem rala duplex, altera cibum è ripis fluminum petit, altera degit in ericeto in locis sylvestribus. Aquaticam illam Coloniæ diu alui, & male uolare deprehendi, & egregiè pugnacem. Rostrum & crura erāt rubra, plumę multis maculis respersę.

Montana verò illa & syluestris crura habet multò breuiora aquatili, & plumas undiq; magis cinereas, sed rubra interim crura habet & rostrum. auis utraq; apud nos regium epulum (real Itali regium vocant, Galli royal, & forte hinc ductum est ralæ vocabulũ. à colore crurum forte erythropus fuerit) vocatur. Hæc ille.

De Cinclo (p. 592).

Turnerus cinclum auem Anglicè interpretatur a water swallow, (quasi dicas hirundinem aquaticam,) Germanicè *ein Steinbeisser*, (sed nostri aliam auẽ, coccothrausten nostrum, *Steinbeisser* appellant.) Auicula (inquit) quam ego cinclum esse puto, galerita paulo maior est, colore in tergo nigro, uentre albo, tibiis longis, & rostro neutiquam breui. Vêre circa ripas fluminum ualde clamosa est & querula, breues & crebros facit uolatus. ¶ Huic Turneri descriptioni cognata uideri potest merula aquatica nostra: magis uerò illa, quam circa Argentoratum *Lyssklicker* appellant: quam non similem modo Turneri cinclo, sed prorsus eandem esse conijcio, cuius figuram sequens pagina cõtinet.

De Motacilla quam nostri albam cognominant (p. 593).

Turnerus in libro de Auibus Cnipológon Aristotelis (id est culicilegã interprete Gaza) hãc auem esse putat....[quotation]....Sed postea in epistola ad me, Culicilegam Aristotelis (inquit) in terra Bergensi uidi, tota cinerei ferè coloris est, & speciem habet pici Martij, illa uerò quam culicilegam esse putabã, est uariola nisi fallor.

De Nycticorace (pp. 602—604).

TURNERUS in litteris ad me missis caprimulgum auem se uidisse scribit prope Bonnam (Germaniæ ciuitatem ad ripam Rheni, supra Coloniam,) ubi à uulgo appellatur *Naghtrauen*, id est coruus nocturnus. Nos in præcedente pagina effigiem adiecimus auis quæ circa Argentoratum, ut audio *Nachtram*, alibi *Nachtrab* nominatur. quæ tamen neq; caprimulgus neq; nycticorax mihi uidetur. [The figure is of the Night-Heron, unmistakably.]

De Onocratalo (pp. 607—608).

Onocrotalus Machliniensis, quæ *Vogelhain* à Brabantis uocatur, quinquaginta annis, ut ipsi ferunt, Machliniæ uixit, cygno maior est. pennę foris albicant, in fundo uerò rubrum

quiddam ostendunt. collum duas spithamas longum est, aut paulò longius, rostrum, quod rubrum habet, dodrantali longitudine est & quatuor uncias longius, & in fine hami propemodum more incuruum & uersus finem latius latiusq, proturbinatur. crura anserinis similia, breuia, nimirum pro magnitudine tanti corporis: in pectore magnum habet ueluti sacculum protuberantem Alis est longissimis, & ipsis in summa extremitate nigris, Guil. Turnerus in epistola ad me....

Bononiæ uisus est mihi onocrotalus (uerba sunt ex epistola Angli cuiusdam amici ad me) plumis cinereis tectus, cygno maior, palmipes, çapite mergi, rostro quatuor palmas ferè longo, & in fine adunco, collo deplumi, amplissimo, ut anatem deuorare posset. Captam aiebãt in lacu Benaco....

Solis uictitat piscibus, & bis tantum anno bibit. Turnerus...Pisces præsertim anguillas auidissimè uorat botaurus auis, Turnerus. idem amicus quidam de onocrotalo ad me scripsit. ego onocrotalum quidem hoc facere non dubito : de botauro dubitari potest, præsertim cum multi etiã non indocti cum onocrotalo eum confundant.

De Perdice (p. 644).

...Quæ Aristoteles & Ouidius de perdice scribunt, omnia nostræ perdici uulgari conueniunt, nempe uolandi nidulandiq, ratio, astutia, circa prolem solicitudo, corporis grauitas, & uocis stridor, à quo etiam nomen accepisse uidetur, Turnerus in epistola ad nos.

The 'Avium præcipuarum...Historia' was reprinted by Dr George Thackeray, Provost of King's College, Cambridge, in 1823; but the reprint is as rare as, if not rarer than, the original. Two copies are in the Library of King's College.

The following is a list of the Birds determined by Turner.

ALAUDIDÆ. *Alauda arvensis.* Lerk or Laverock. German *Lerch*, p. 80.

A. arborea. Wodlerck, p. 80.

A. sp.? Wilde Lerc or Heth Lerk. G. *Heid Lerch*, p. 80.

Galerita cristata. [No English name.] G. *Copera*, p. 80.

ALCEDINIDÆ. *Alcedo ispida.* Kynges fissher. G. *Eissvogel*, pp. 18—22. [Turner recognised two kinds of Kingfishers described by Aristotle and Pliny, but does not state what they are].

ANATIDÆ. *Anas boscas.* Duck, pp. 22, 48.
Anser (2 species.) Gose. G. *Ganss*, p. 22.
Bernicla leucopsis. Brant or Bernicle Gose, p. 26.
Cygnus olor. Swan. G. *Swän*, p. 120.
Mareca penelope. Wigene, p. 48.
Nyroca ferina. Pochard, p. 48.
Querquedula crecca. Tele, p. 48.
Tadorna cornuta. Bergander, p. 24.

ARDEIDÆ. *Ardea* sp.? (white). Cryel or Dwarf Heron, p. 38.
A. cinerea. Heron. G. *Reyger*, p. 36.
Botaurus stellaris. Bittour, Buttor, Buttour, or Myre Dromble. G. *Pittour*, *Rosdom*, pp. 38, 40, 122.

CAPRIMULGIDÆ. *Caprimulgus europæus.* [No English name], p. 48.

CERTHIIDÆ. *Certhia familiaris.* Creper, p. 52.

CHARADRIIDÆ. *Charadrius pluvialis.* Pluver. G. *Pulver*, p. 132.
Vanellus vulgaris. Lapwing. G. *Kywit*, pp. 76, 174.

CICONIIDÆ. *Ciconia alba.* Stork. G. *Storck*, Sax. *Ebeher*, p. 54.

CINCLIDÆ. *Cinclus aquaticus.* Water-Craw, p. 22.

COLUMBIDÆ. *Columba* sp.? Dove. G. *Taube*, Sax. *Duve*, p. 59. [Venice Dove, p. 62.]
C. œnas. Stocdove. G. *Holtztaube*, p. 60.
C. palumbus. Coushot or Ringged Dove. G. *Ringel Taube*, p. 60.
Turtur communis. Turtel Duve, p. 60.

CORVIDÆ. *Corvus corax.* Raven. G. *Rabe*, p. 64.
C. cornix. Winter Crow, p. 64.
C. corone. Crow. G. *Krae*, *Kraeg*, p. 64.
C. frugilegus. [No English name], p. 64.
C. monedula. Caddo, Chogh, Ka. G. *Döl*, Sax. *Älke*, p. 92.
Garrulus glandarius. Jay. G. *Mercolphus*, p. 144.
Nucifraga caryocatactes. [No English name.] G. *Nousbrecher*, p. 94.
Pica rustica. Py, Piot. G. *Elster*, *Atzel*, pp. 142, 144.
Pyrrhocorax graculus. Cornish Choghe. G. *Bergdöl*, p. 90. [Confounded with *P. alpinus.*]

CUCULIDÆ. *Cuculus canorus.* Cukkow, or Gouke. G. *Kukkuck*, p. 66.

CYPSELIDÆ. *Cypselus apus.* Chirche Martnette. Rok Martinette. G. Kirch Swalbe, pp. 100, 102.
C. melba. Great Swallow. G. *Geyr Swalbe*, p. 102.

FALCONIDÆ. *Accipiter nisus.* [No English name], p. 66.
Aquila sp.? Right Egle. G. *Edel Ärn*, p. 36.
Astur palumbarius [?] Sparhauc. G. Sperwer, p. 18.
Buteo vulgaris. Bushard, p. 16.
Circus æruginosus. Balbushard, p. 32.
C. cyaneus. Hen-Harroer, Ringtale, p. 18.
[Turner calls the male Hen-Harroer, the female Ringtale, erroneously considering them two species.]
Falco æsalon. Merlin. G. *Smerl*, p. 16.

FALCONIDÆ. *F. subbuteo.* Hobby, p. 18.
Gypaëtus barbatus [?]. [No English name], p. 128.
Haliaëtus albicilla. Egle or Erne. G. *Ärn*, *Adler*, p. 30.
Milvus ater. [No English name], p. 116.
M. ictinus. Glede, Puttok, or Kyte. G. *Weye*, p. 116.
Tinnunculus alaudarius. Kastrel, Kistrel, or Steingall, p. 166.

FRINGILLIDÆ. *Carduelis elegans.* Goldfinche. G. *Distelfinck*, *Stigelitz*, pp. 40, 50.
C. spinus. Siskin. G. *Zeysich*, *Engelchen*, p. 108 (cf. p. 50).
Emberiza citrinella. Yelowham, Yowlryng. G. *Geelgorst*, p. 106.
E. miliaria. Bunting. G. *Gersthammer*, pp. 134, 158.
E. schœniclus. Rede Sparrow. G. *Reydt Müss*, pp. 102, 134.
Fringilla cœlebs. Chaffinche, Sheld-appel, Spink. G. *Bůchfink*, p. 72.
F. montifringilla. Bramlyng. G. *Rowert*, p. 72.
Ligurinus chloris. Grenefinche. G. *Kirsfincke*, pp. 104, 106.
Linota cannabina? Linot. G. *Flasfinc*, pp. 50, 158.
Passer domesticus. Sparrow. G. *Můsche*, *Spätz*, *Lüningk*, Sax. *Sperlingk*, p. 132.
Pyrrhula europœa. Bulfinche. G. *Blödtfinck*, p. 160.
Serinus canarius. Canary Bird, p. 108.

GRUIDÆ. *Grus communis.* Crane. G. *Krän*, *Kränich*, pp. 94, 96.

HIRUNDINIDÆ. *Cotile riparia.* Bank Martnet. G. *Über Swalbe*, *Speiren*, p. 102.
Hirundo rustica. Swallowe. G. *Schwalb*, Sax. *Swale*, pp. 96, 100, 102.

IBIDIDÆ. *Comatibis eremita.* [Redcheeked Ibis]. G. *Waltrap*, pp. 92, 94.

LANIIDÆ. *Lanius excubitor.* Schric, Shrike, or Nyn Murder. G. *Nuin Mürder*, *Neun Mürder*, pp. 116, 118, 168.
Lanius minor? [No English name], p. 168.

LARIDÆ. *Hydrochelidon nigra.* Stern, p. 78.
Larus sp.? [Grey Gull.] Se Cob or See Gell, p. 78.
L. sp.? [White Gull.] White Semaw. Se Cob or Seegell. G. *Wyss mewe*, pp. 74, 78.
L. ridibundus. White Semaw with a blak cop, pp. 74, 76.

MEROPIDÆ. *Merops apiaster.* [No English name], p. 112.

MOTACILLIDÆ. *Anthus pratensis.* Titlyng, p. 68.
Motacilla lugubris or *M. alba.* Wagtale. G. *Wasser Steltz*, *Quikstertz*, p. 64.

ORIOLIDÆ. *Oriolus galbula.* Witwol. G. *Witwol*, *Weidwail*, *Kersenrife*, pp. 148, 172, 174.

OTIDIDÆ. *Otis tarda.* Bistard or Bustard. G. *Träp* or *Trap Ganss*, pp. 130, 166.

PANDIONIDÆ. *Pandion haliaëtus.* Osprey. G. *Vishärn*, pp. 34, 36.

PARIDÆ. *Parus cœruleus.* Non. p. 132.

PARIDÆ. *P. major.* Great Titmous, or Great Oxei. G. *Kölmeyse*, p. 130.

P. palustris or *P. ater.* Less Titmous. G. *Meelmeyse*, p. 130.

PHALACROCORACIDÆ. *Phalacrocorax carbo.* Cormorant. G. *Důcher*, p. 110.

P. graculus? Douker (pt), Loun, *Důcher* (pt.), p. 176.

PHASIANIDÆ. *Attagen* [possibly *Bonasa sylvestris*, the Hazel Grouse], pp. 42, 44.

Gallus ferrugineus (*domesticus*). Cok, Hen. G. *Hän*, *Hen*, Sax. *Hön*, p. 82.

Numida meleagris. [No English name?] Kok of Inde? pp. 82, 86, 140.

Pavo cristatus. Pecok. G. *Pffaw*, Sax. *Pageliin*, p. 136.

Perdix cinerea. Pertrige. G. *Velt hön*, *Raphön*, p. 138.

Phasianus colchicus. Phesan. G. *Fasant*, *Fasian*, p. 140.

PHŒNIX. [No English name], p. 140.

PICIDÆ. *Dendrocopus major?* Specht, or Wodspecht. G. *Elsterspecht*, pp. 146, 148.

Gecinus viridis. Hewhole, Huhol, Raynbird? G. *Grünspecht*, pp. 88, 112, 114, 146, 148.

Iynx torquilla. [No English name], pp. 146, 148.

Picus martius. [No English name]. G. *Craspecht*, p. 148.

PLATALEIDÆ. *Platalea leucorodia.* Shovelard. G. *Lefler*, *Löffel Ganss*, pp. 38, 150.

PODICIPEDIDÆ. *Podicipes minor*, Douker (pt), *Důcher* (pt), p. 176.

PROCELLARIIDÆ. *Puffinus* sp.? Bird of Diomede, p. 70.

PSITTACIDÆ. Parrot. Popinjay. G. *Papegay*, p. 150.

RALLIDÆ. *Crex pratensis.* Daker Hen, Rale (?) G. *Schryk*, *Scrica*, pp. 70, 128, 140.

Fulica nigra. Cout, pp. 32, 76.

Gallinula chloropus. Mot Hen or Water Hen. G. *Wasser Hen*, p. 170.

Porphyrio cælestis. [No English name], p. 152.

SCOLOPACIDÆ. *Actitis hypoleuca.* Water Swallow. G. *Steynbisser*, pp. 54, 56.

Limosa belgica. Godwitt or Fedoa, p. 44.

Scolopax rusticula. Wodcok. G. *Holtz Snepff*, pp. 42, 86.

Totanus calidris. Redshanc, p. 102.

SITTIDÆ. *Sitta cæsia.* Nut-jobber. G. *Nushäkker*, *Meyspecht*, p. 162.

STRIGIDÆ. *Asio otus.* Horn Oul. G. *Ranseul*, *Schleier Eul*, p. 130.

Bubo ignavus. Lyke Foule. G. *Schuffauss*, *Schüffel*, *Kautz*, p. 46.

Strix stridula (?) Owl, Howlet. G. *Eul*, Sax. *Üle*, p. 120.

STRUTHIONIDÆ. *Struthio camelus.* Oistris. G. *Strauss*, p. 164.

STURNIDÆ. *Sturnus vulgaris.* Sterlyng. G. *Stär*, *Stör*, p. 164.

SULIDÆ. *Sula bassana.* Solend Guse, p. 28.

SYLVIIDÆ. *Accentor modularis* (?) Hedge-sparrow, or Dike Smouler. G. *Grassmusch*, *Koelmussh*, p. 136.

Daulias luscinia. Nyghtyngall. G. *Nachtgäl*, p. 108.

SYLVIIDÆ. *Erithacus rubecula.* Robin Redbreste. G. *Rötbrust*, *Rötkelchen*, p. 154.

Pratincola rubicola. Stonchatter or Mortetter. G. *Klein Brachvogelchen*, p. 158.

Regulus cristatus (?). [No English name.] G. *Gold Hendlin*, pp. 154, 168.

Ruticilla phœnicurus. Rede Tale. G. *Rötstertz*, p. 154.

Saxicola œnanthe. Arlyng, Clotburd, Smatche or Steinchek. G. *Brechvögel*, p. 52.

Sylvia atricapilla (?). [No English name.] G. *Grasmuklen*, p. 44.

S. rufa? Lingett. G. *Graesmusch*, *Grassmusch*, p. 111, [cf. p. 136].

TETRAONIDÆ. *Coturnix communis.* Quale. G. *Wachtel*, p. 62.

Lagopus mutus. [No English name], p. 104.

L. scoticus. (?) Morhen, p. 86.

Tetrao tetrix. [No English name], p. 42.

TROGLODYTIDÆ. *Troglodytes parvulus.* Wren. G. *Kuningsgen*, *Zaunküningk*, p. 152.

TURDIDÆ. *Turdus iliacus*, *musicus*, *viscivorus.* Thrusche, Thrushe, Throssel, Mavis, or Wyngthrushe. G. *Drossel*, *Durstel*, *Weingaerdsvoegel*, pp. 170, 172.

T. merula. Blak Osel or Blakbyrd. G. *Merl*, *Amsel*, p. 114.

T. pilaris. Feldfare or Feldefare. G. *Krammesvögel*, *Wachholtervögel*, pp. 58, 170. [Confounded with Mistletoe Thrush?]

UPUPIDÆ. *Upupa epops.* Howpe. G. *Houp*, *Widhopff*, p. 174.

VULTURIDÆ. *Vultur* sp. Geir. G. *Geyr*, p. 176.

AVIVM

PRAECIPV ARVM, QVARVM

APVD PLINIVM ET ARI-
ſtotelem mentio eſt, breuis &
ſuccincta hiſtoria.

*Ex optimis quibuſque ſcripto-
ribus contexta, ſcholio illu
ſtrata & aucta.*

*Adiectis nominibus Græcis, Germanicis &
Britannicis.*

*Per Dn. Guilielmum Turnerum, artium & Me-
dicinæ doctorem.*

*Coloniæ excudebat Ioan. Gymnicus,
Anno M. D. XLIIII.*

DE HISTORIA AVIUM.

[p. 3] Illuſtriſſimo VValliæ principi, Eduuardo filio hæredi, ſereniſſimi & potentiſſimi Henrici VIII. regis Angliæ, Franciæ, & Hiberniæ, Guilielmus Turnerus S. P. D.

PRUDENS admodum, &, ſi quid ego intelligo (illuſtriſſime princeps) neceſſaria imprimis regij prophetæ fuit admonitio, qua reges, principes & iudices terræ, ut intelligerent, & eruditionem conſequerentur, admonuit. Nam ut ſummus ille rerũ architectus Deus optimus maximusq́ꝫ, caput ſuper omnes reliquas corporis partes in homine, qui perfectiſſimę in ſe [p. 4] Reipublicæ ſimulachrum gerit, collocauit, & omnes quinqꝫ ſenſus ſimul in eo repoſuit, ut pro reliquis omnibus mẽbris (quibus ſolus tactus eſt conceſſus) uideret, audiret, guſtaret & odoret, & eorum ſaluti conſuleret : ita principem Reipublicę, corpori ex multis membris conflato, ueluti caput præfecit, ut prudentia, eruditione, & ſenſibus ſuis non tam exterioribus, quã interioribus, totius Reipublicæ commodis & ſaluti proſpiceret. In pedibus uiſum, in tibijs auditum, in manibus olfactum, in brachijs odoratum nemo requirit: ſed hæc omnia in capite requiruntur.

To the most illustrious Prince of Wales, Edward, son and heir of the most serene and mighty Henry VIII, King of England, France and Ireland, William Turner wishes long life and health.

EXCEEDING wise, and if I understand aright, necessary above all things, most illustrious Prince, was the warning of the royal prophet, in which he admonished kings, princes, and judges of the earth that they should have understanding and seek learning. For, as that architect supreme of the universe, God most good and great, placed the head above all the remaining parts of the body in man, who in himself shews forth the image of a most perfect State, and stored up in it all the five senses at once, that it should see, hear, taste, and smell for all the remaining members (to which touch alone has been allowed), and should consult for their well-being; so he hath set the Prince, as it were a head, over the State, a body welded together of many members, that he should provide for the advantage and well-being of the whole State by his wisdom, learning and senses, not so much external as internal. No one demands sight in the feet, hearing in the legs, smell[1] in the hands, or smell in the arms; but all these things are necessary in the head.

[1] This should probably be "taste" (gustum).

Quum igitur tot ſenſus in capite uni tantũ corpori prẹfecto requirantur: quot ſenſus, quantum ſapientiæ & eruditionis ab eo capite exi-
[p. 5] guntur, cui plus quàm trecentorum milium corporum præfectura committitur? Quòd ſi quis forſan reſpondeat, non in principe, ſed in ijs ſolis, qui illi à conſilijs ſunt, eruditionem & prudentiam requiri: hunc ego dignum cenſerẽ, qui pro tali reſponſo, omnibus ſenſibus, excepto tactu, orbatus in media ſylua uepribus & ſpinis denſa, caueis & foſſis formidabili, quatuor ducibus comitatus ſtatueretur, nobis dicturus, nũ proprijs malit uti ſenſibus an alienis? & num tutius illi ſit, ducum ſuorum incertorum ſenſibus, an proprijs duci? & qua ratione cæcus & ſurdus odoratu & guſtu deſtitutus ipſe, cẹci'ne an uidẽtes ſui ſint duces, dignoſcere poſſit?

In conſiliarijs ſummam prudentiam & eru-
[p. 6] ditionem non uulgarem requiri, non diffiteor: uerùm non in ijs ſolis, nam ſi illi, qui principi ſunt à conſilijs, ad tempus bene conſulant, & poſtea in ipſius perniciem malè ſuadeant, ut Abſaloni Achitofelem feciſſe legimus: quomodo pernicioſum illorum conſiliũ ipſe ſubodorabitur & depræhendet, niſi eruditione & prudentia conſiliarios ſuos aut ſuperet, aut ſaltem æquet? Quare nõ in conſiliarijs tantùm, ſed in principe ipſo eruditio & ſapiẽtia requiruntur. Non deſunt, qui ſatis eſſe principi exiſtimant, quo cæteris mortalibus præſtet, ſi regio ueſtitu, diuitijs, copijs, ſcitè pulſando teſtudinem, & tela dextrè uibrando, ſubditis ſuis prẹluceat: uerùm fortiſ-
[p. 7] ſimi quiqꝫ & ſapientiſſimi reges longè diuerſum

Inasmuch therefore as so many senses are requisite in the head, which is set over one body alone, how many senses and what a wealth of wisdom and learning are demanded from that head, to whom more than three hundred thousand bodies are given in charge? But if any should chance to answer that learning and wisdom are needed not in the Prince, but only in those who are his councillors, I should consider it fitting that he for such a reply should be set, accompanied by four guides, in the midst of a wood tangled with briers and thorns, and dangerous with its pits and ditches, deprived of all his senses, except that of touch, and should tell us whether he preferred to use his own senses or those of others: or whether it would be safer for him to be led by the senses of his doubting guides or by his own; and in what way he, being blind and deaf, and destitute of smell and taste, could determine whether his guides were blind or able to see.

I fail not to confess that the highest wisdom, and learning of no common sort, are requisite in councillors, but not in them alone; for if they who are the advisers of the Prince, give good counsel for the time, and afterwards prompt him ill to his destruction, as we read that Achitofel did in the case of Absalom, how shall he smell out and detect their fatal advice, unless he either excels or at least equals his councillors in learning and wisdom? Wherefore not only in councillors but in the Prince himself are learning and wisdom requisite. There are not wanting those who think it enough for a Prince, as matters in which he should surpass other mortals, if he outshines his subjects in royal garb, in riches, in resources, in cunningly striking the lyre, and in skilfully throwing the spear; but all the bravest and wisest kings have

ſenſerunt. Nã Mithridates rex Põti & Bithyniȩ, ſe regnorum ſuorum caput eſſe intelligẽs, et tot corporibus, quot prȩerat, unicã uernaculã ſuam linguã minimè ſufficere, uiginti duas linguas gẽtium, quas ſub ditione ſua habuit, ita perfectè didicit & percalluit, ut uiginti illarum gentium uiris ſine interprete promptè reſponderit, & ſua cuique lingua non ſecus atque gentilis fuiſſet, locutus fuerit. Idẽ rerum abditas naturas ita perueſtigauit, & in re medica ita fœliciter fuit uerſatus, ut aduerſùs lethalia uenena antidotum, quod hodie etiamnum ab eo nomen ſortitum, Mithridatium appellatur, ſuo Marte inuenerit. Alexander ille Macedonum rex, tam [p. 8] naturæ quàm fortunæ dotibus iure ſuſpiciendus, tanto bonarum artium & philoſophiæ potiſſimum ſtudio flagrauit, ut etiã in zelotypiam quandam literariam inciderit. Nam is cùm omnem propè Aſiam armis & exercitu teneret, ubi primũ Ariſtotelem libros ſuos de auscultatione phyſica inuulgaſſe acceperat, in tantis negocijs cum Ariſtotele, miſſa ſtatim epiſtola de editis libris, his uerbis expoſtulabat: Quòd diſciplinas ἀκροαματικὰς edidiſti, non rectè feciſti. nam qua alia re cæteris præſtare poterimus, ſi ea, quæ abs te accepimus, omnium prorſus fuerint communia? Quippe ego doctrina anteire malim, quàm copijs atq; opulentijs. Hæc Alexander.

Diuino approbatus oraculo rex ille Dauid,

thought quite differently. For Mithridates, king of Pontus and Bithynia, understanding that he was the head of his domains, and that his native tongue alone was by no means sufficient for the numerous bodies, over which he reigned, learned so perfectly and understood so thoroughly the twenty-two tongues[1] of the nations, which he had under his sway, that he gave immediate answers to twenty men of those nations without an interpreter, and spoke to each in his own tongue just as if it had been native to him. He also so thoroughly traced out the hidden natures of things, and occupied himself to such good purpose in the science of medicine, that he discovered by his own exertions an antidote to deadly poisons, which even to-day is called Mithridatium, a name derived from him. The great Alexander, king of the Macedonians, rightly renowned as much for the gifts of nature as for those of fortune, burned with so great a zeal for the noble arts, and philosophy in particular, that he even descended to a sort of literary jealousy. For though he was holding almost all Asia by force of arms and his troops, when first he heard that Aristotle had made public his books 'De Auscultatione Physica,' in the midst of such great concerns he expostulated with Aristotle in the following words, a letter having been at once sent off concerning the publication of the books: "In that you have published your teachings called ἀκροαματικαὶ you have not done rightly; for in what other thing shall I be able to excel the rest, if those things, which I have heard from you, become henceforth the common property of all? For I should prefer to stand first in learning rather than in resources and wealth." Thus said Alexander.

The great king David, approved by the voice of

[1] 'Duas' is perhaps a misprint for 'duarum.'

[p. 9] qui & propheta fuit diuino numine adflatus, atq; ideo qd regi maximè neceſſariũ foret, cognoſcens, ante omnia literas, nempe ſacras expetiuit, ut ſibi tẽperare non potuerit, quin diceret, Benedictus es domine, doce me iuſtificationes tuas, in uia mandatorũ tuorũ delectatus ſum, ſicut in omnibus diuitijs : in mandatis tuis exercebor, & conſiderabo uias tuas. Reuela oculos meos, & conſiderabo mirabilia de lege tua. Bonitatẽ & diſciplinã & ſcientiam doce me : ego autem in toto corde meo ſcrutabor mandata tua. Niſi quòd lex tua meditatio mea eſt, tunc fortè periſſem in humilitate mea. Quàm dulcia faucibus meis eloquia tua, ſuper mel ori meo. Bonum mihi lex oris tui ſuper milia auri & argẽti. [p. 10] Lucerna pedibus meis uerbum tuum, & lumen ſemitis meis. Declaratio ſermonum tuorum illuminat, & intellectum dat paruulis. Hactenus rex Dauid, & pace & bello omnium regum illuſtriſſimus.

Rex Solomon huius filius, omniũ, quos unquã terra genuit, ſapientiſs. cuius unius autoritati plus tribuendum eſt, quàm ſexcentis adulatoribus diuerſum ſuadẽtibus, cùm totius orbis conditor & omnium bonorum largitor Deus pater, illi, quod ſibi optimum, & ex uſu ſuo maximè fore iudicaret, ultro offerret, & poſcenti mox ſe daturũ promitteret, ad hunc modum, ut diuinæ literæ teſtantur, reſpondit. Nunc domine Deus, tu me regnare feciſti ſeruum tuum pro [p. 11] Dauide patre meo, ego autem ſum puer paruulus, & ignorans ingreſſũ, & introitũ meũ : & ſeruus tuus in medio eſt populi, quem elegiſti,

God, who was moreover a prophet filled with divine inspiration, and therefore well aware of what was especially necessary for a king, sought before all things learning, and that of course divine, so that he was unable to restrain himself from saying "Blessed art thou, O Lord, teach me thy righteousness, I have delighted in the way of thy commandments, as in all riches: in thy statutes will I exercise myself, and I will consider thy ways. Open thou mine eyes, and I will consider the wonderful things of thy law. Teach me goodness and instruction and learning; but with my whole heart will I examine thy commandments. Unless thy law had been my meditation, then should I perchance have perished in my low estate. How sweet are thy sayings to my mouth, better than honey to my lips. The law of thy mouth is a good to me beyond thousands of gold and silver. Thy word is a lantern unto my feet, and a light unto my paths. The telling of thy discourses giveth light and understanding to babes." Thus far king David, the most illustrious of all kings both in peace and war.

King Solomon, his son, the wisest of all that earth ever bore, to whose single authority more weight is to be given than to six hundred flatterers persuading to a different course, when God the Father, maker of all the world and giver of all good things, of his own accord offered to him what he should judge to be best for himself and for his greatest advantage, and promised that he would grant it at once on his request, replied in this manner, as the Scriptures testify. "Now, O Lord God, thou hast made me, thy servant, to reign in the room of David my father, but I am a little child, and know not my coming in and entering; and thy servant is in the midst of the people whom thou hast chosen, an in-

populi infiniti, qui numerari & ſupputari non poteſt prę multitudine. Dabis ergò ſeruo tuo cor docile, ut populum tuum iudicare poſſit, & diſcernere inter bonum & malum: quis enim poteſt iudicare populum iſtum, populum tuum hunc multum? Huc uſq; Solomon, qui in philoſophia tam diuina quàm humana ita non multis pòſt annis profecit, ut de ſtirpibus à cedro uſq; ad hiſſopum diſputauerit, & de beſtijs, uolucribus, reptilibus, & piſcibus diſſeruerit.

Quare, prudentiſſimi quique principes, & [p. 12] fortiſſimi, nõ ſatis habebant, ſubditos ſuos diuitijs, honoribus, ueſtitu, inceſſu, & bellica gloria excellere, niſi literis, linguis, philoſophia tam diuina quàm humana inſuper multùm ſuperarent, & à tergo relinquerẽt. Quod pater tuus omnium regum, qui hodie uiuunt, eruditiſſimus, ſatis ut regem tantum decet, intelligens, & cui Reipub. gubernaculum committitur, quàm neceſſaria literæ & philoſophia ſint, prudenter ſecum perpendens, liberos ſuos ſemper eruditiſſimis quibuſque præceptoribus commiſit. Duci enim Richmundiæ, piæ memoriæ, fratri tuo Georgium Folberium præceptorẽ olim meum, uirum inſigniter doctum, et mirum rectè inſtituendæ iuuentutis artificem, & tibi nũc uirum longè [p. 13] doctiſſimum (uti audio) præfecit.

Qua de cauſa, illuſtriſſime & optime princeps, ſapiẽtiſſimorum & fortiſſimorum regum exempla ſecutus, atq; potentiſſimi & eruditiſſimi patris tui conſilio obtemperans, qui te ad meliores imbibendas literas, nunquam non inuitat, incitat & hortatur, dum ætas tua adhuc tenera

numerable people, a people which cannot be numbered or counted for their multitude. Thou shalt give therefore to thy servant a heart that may be taught, that he may be able to judge thy people, and to discern between good and evil: for who is able to judge this people, this great people of thine?" Thus far spoke Solomon, who not many years afterwards so excelled in philosophy both divine and human that he disserted about plants from the cedar even to the hyssop, and discoursed of beasts, birds, reptiles, and fishes.

Wherefore all the most wise and brave Princes have not considered it sufficient to surpass their subjects in riches, honours, garb, gait, and warlike glory, unless beyond this they excelled them far in learning, tongues, and philosophy both divine and human, and left them in the rear. And this your father, the most learned of all the kings who are alive at the present day, well understanding, as becomes so great a king, and one to whom the helm of the State is entrusted, wisely pondering in his mind how necessary learning and philosophy are, always committed his children to the care of the most learned of instructors. For over your brother the Duke of Richmond, of pious memory, he set Georgius Folberius, once my tutor, a man of remarkable learning, and a wondrous handicraftsman for rightly instructing youth, and now over you (as I hear) a man by far the most learned of all.

Wherefore, most illustrious and worthy Prince, following the steps of the wisest and bravest kings, and yielding to the advice of your most powerful and learned father, who so constantly invites you to the draught of superior learning, spurs you on, and exhorts you, while your years are yet tender and

eſt, & literarũ capaciſſima, omne genus bonarum literarum obuijs ulnis amplectere, diſce, & imbibe, & exantlati in bonas literas laboris olim te minimè pœnitebit. Sed ut ad propoſitam metam minori cum negocio poſſis peruenire, libellum De hiſtoria auium, in quo Latinis nominibus Græca, Germanica & Britãnica in [p. 14] gratiam tuam appoſui, ex Ariſtotele & Plinio, & optimis quibusq; ſcriptoribus contexui. Hunc ego nominis tui celebritati dedico, & dono: etiã atq; etiã te obteſtans, ut hoc meum qualecunq; munuſculum æqui boni'q; conſulas. Quod ſi te factur̄um intellexero, & hunc libellum figuris & auium moribus, & medicinis auctum, & de herbis alium etiam librum, breui, uolente Deo, in lucem emittam. Vale. Dominus Ieſus te nobis ſanctiſſimis moribus inſtitutum, & optimis literis imbutum, quàm diutiſſimè inculumem conſeruet.

Coloniæ 5. Idus Februarij, Anno M.D.XLIIII.

most amenable to learning, embrace with open arms every kind of noble literature, learn and drink it in, and hereafter you will surely not repent of the labour expended upon this noble literature. But that you may be able to reach with less trouble the goal that is laid before you I have compiled from Aristotle and Pliny and all the best writers this little book on 'The History of Birds,' in which I have placed for your pleasure the Greek, German, and British names side by side with the Latin. This I dedicate and offer to the glory of your name: again and again praying you to receive this little gift, such as it is, with fair and favourable consideration. And if I understand that you will do this, I will shortly, God willing, bring to the light of day a further edition of this little book with figures of the birds, their habits, and curative properties, as well as another book on plants. Farewell. May the Lord Jesus preserve you as long as possible unharmed to us, trained in most holy ways and filled with the best of learning.

Cullen [Cologne]. February 9th, 1544.

[p. 15] De decem generibus Accipitrum.

ARISTOTELES[1].

Buteo.

Aeſalo.
Circus.

Percæ
Fringillarij.
Rubetarij.

ACCIPITRUM genus præcipuum Buteo eſt, Triorcha[2] à numero teſtium nuncupatus: ſecundum æſalo, tertium circus. Stellaris autem, palumbarius, & pernix[3] differunt. Appellantur ſubuteones, qui latiores[4] ſunt: alij percæ & fringillarij uocantur: alij læues[5] & rubetarij, qui abundè uiuunt[6], atque humiuolę ſunt. Genera non pauciora quàm decem eſſe accipitrum aliqui prodiderunt, quæ modo quoqꝫ uenandi[7] inter ſe diſſident. Alij enim columbam humi conſidentem, rapiunt, uolantem non appetunt: alij ſuper arborem, aut tale quid conſcendentem, uenantur: ſin humi est, aut uolat, [p. 16] non inuadunt. Alij neqꝫ humi, neqꝫ in ſublimi manentẽ, adgrediũtur, ſed uolantem capere conantur. Fertur etiam à columbis quodqꝫ accipitrũ genus cognoſci. Itaqꝫ cùm accipiter prouolat, ſi ſublimipeta eſt, manent quo conſtiterunt loco: ſed ſi humipeta qui prouolat, eſt, non manẽt, ſed continuò auolant.

[1] *Hist. An.* Bk IX. 128—130.
[2] Aristotle has simply κράτιστος μὲν ὁ τριόρχης.
[3] Other readings are πτερνὶς, πέρνης, πτέρνης.
[4] Instead of πλατύτεροι, some texts have πλατύπτεροι, which would make better sense and mean 'broad-winged.'
[5] λεῖοι, or according to another text ἐλειοί.
[6] The word εὐβιώτατοι here, and corresponding expressions throughout the passages quoted in this book, might possibly mean that the birds in question have no particular faults, or are of ordinary respectability. Gaza, however, followed as usual by Turner, seems to have interpreted the word rightly here.
[7] These three words are not found in Aristotle.

Of the ten kinds of Accipitres.

ARISTOTLE.

THE chief kind of Accipitres is Buteo, which from the number of its testicles is named Triorcha, Æsalo is the second, Circus is the third. Again Stellaris, Palumbarius, and Pernix differ. Those which have more breadth are called Subuteones; other kinds are named Percæ and Fringillarii; others Læves and Rubetarii, which get their living most easily, and fly near to the ground. Some have asserted that there are no fewer than ten kinds of the Accipitres which differ from each other in their several modes of hunting. For some sorts seize a Dove when sitting on the ground, but do not touch one flying; others seek their prey when perched upon a tree, or such like, but if it be on the ground or flying do not attack it. And others seize it neither on the ground, nor when resting aloft, but strive to catch it flying. Moreover it is said that each kind of Accipitres is recognised by Doves. So, when the Accipiter comes forth, if it be such as hunts on high, they stay where they have settled, but, if that which comes be such as takes them on the ground, they stay not, but forthwith fly off.

PLINIUS[1].

Circus. †pecuariȩ Buteo. Cymindis. [p. 17]

Accipitrum genera ſedecim inuenimus. Ex ijs circon claudum altero pede, proſperrimi augurij nuptialibus negocijs, & †pecuniariæ rei[2]. Triorchen à numero teſtiũ, cui principatum in augurijs Phœmone dedit: buteonẽ hunc appellant Rom. Aeſalona Græci uocant, qui ſolus omni tẽpore apparet. Cæteri hyeme abeunt. Nocturnus accipiter cymindis uocatur, rarus etiam in ſyluis, interdiu minùs cernens: bellum internecinum cum aquila gerit: cohærentesq; ſæpè præhenduntur. Hæc Plinius.

Quanquam Ariſtoteles decem eſſe accipitrum genera tradat, & Plinius ſedecim: neuter tamen horum hæc ita diſtinxit genera, & deſcripſit, ut procliue ſit lectori ſuum cuique peculiare nomen ex illorum præſcriptis imponere. Quare à me nemo horum exactam differentiam, & cuiuſque nomen Britannicum aut Germanicum cum Latino & Græco coniunctum, iure poterit exigere. Ego tamen, quod nomen Britannicum, cuiq; Latino imponẽdum eſſe cenſeo, lectorem minimè celabo.

Buteo. *Buteo* τριόρχης *Græcè dictus, Anglorum busharda eſt, niſi fallar: nam miluo magnitudine æquiparatur, ſemperq; ipſe cernitur, qualem Ariſtoteles octauo libro de hiſtoria animalium buteonem deſcribit.*

Aeſalo. Αισάλων, *quoniam iuxta Plinij ſententiam omni tempore apparet, & inter minores accipitres ſola merlina ſiue ſmerla, ſemper adpareat, mihi Anglorum merlina, & Germanor. ſmerla eſſe uidetur.*

[1] *Hist. Nat.* Lib. x. cap. viii.

[2] If the reading pecuariæ is accepted, the meaning would appear to be 'for cattle breeding.'

PLINY.

Of Accipitres we have found sixteen kinds. Circus among them, halting in a foot, of lucky omen in nuptial affairs and money business. Triorches next, to which Phœmone[1] gave the foremost place in auspices, named from the number of its testicles: the Romans call it Buteo and the Greeks Æsalon: it is the only kind which may be seen at every time. The rest leave us in winter. An Accipiter that flies by night is called Cymindis; it is rarely found in woodlands, in the day it scarce can see: it wages deadly warfare with the Aquila, and they are often captured clinging to each other. So far Pliny.

Though Aristotle may set forth that there are ten kinds of Accipitres, and Pliny that there are sixteen, yet neither of them has distinguished or described the kinds so that it may be easy for a reader to apply to each its proper name from their accounts. So no one can in fairness claim from me their exact difference, nor yet the British or the German name of each, together with the Latin or the Greek equivalent. I will, however, surely not conceal from you, my reader, what I think to be the British name, and to which Latin name it ought to be applied.

Buteo, called in Greek *τριόρχης*, if I do not err, is the Buzzard of the English, for it is compared with Milvus as to size; moreover it is seen at all times, and is such a bird as Aristotle makes his Buteo in the eighth book of the 'History of Animals.'

Αἰσάλων, since in Pliny's judgment it appears at every season, and among the smaller Hawks the Merlin or the Smerl alone seems to appear[2] at all times, is, I think, the Merlin of the English and Smerl of the Germans.

[1] Phœmone, called 'Daughter of Apollo,' was a priestess at Delphi. (See Pliny *Hist. Nat.* ed. Hardouin: Lipsiæ, 1791, Index Auctorum, p. 340.)

[2] This seems to be the force of the subjunctive here, if it is not an oversight.

[p. 18] Palumbarius. *Accipitrem palumbarium ideo Anglorum ſparhaucam, & Germanorum ſperuuerum eſſe puto, quòd palumbes, columbos, perdices & grandiuſculas aues inſequatur.*

Fringillarius. *Fringillarium Anglorum hobbiã eſſe conijcio. Eſt autem hobbia accipiter minimus, coloris cæteris nigrioris. In capite duos habet nigerrimos in pallido neuos. Galeritas & fringillas plerumq; captat, in excelſis arboribus nidulatur, & hyeme nuſquam cernitur.*

Rubetarius. *Rubetarium eſſe credo accipitrem illum, quem Angli hen harroer nominant. Porrò ille apud noſtros à dilaniandis gallinis nomen habet. Palumbarium magnitudine ſuperat, & coloris eſt cinerij. Humi ſedentes aues in agris, & gallinas in oppidis & pagis repentè adoritur. Præda fruſtratus, tacitus diſcedit, nec unquam ſecundum facit inſultum. Hic per humum omnium uolat maximè.*

Subbuteo. *Subbuteonem eſſe puto, quem Angli ringtalum appellant, ab albo circulo, qui caudam circuit. Colore eſt medio inter fuluum & nigrū, buteone paulò minor, ſed multò agilior. Prædam eodem modo, quo ſuperior captat.*

[p. 19]

DE ALCEDONE.

Ἀλκυών, *alcedo, Anglicè the kynges fiſsher, Germanicè* eyn eißuogel.

Aristoteles[1].

Alcedo non multò amplior paſſere est, colore tum uiridi, tum cœruleo, tum etiam leuiter purpureo inſignis: uidelicet non particulatim colore ita diſtincta, ſed ex indiſcreto uariè refulgens corpore toto & alis & collo, roſtrum ſubuiride, longum & tenue. Alcedonum[2] quo-

[1] *Hist. An.* Bk IX. 85.

[2] *Hist. An.* Bk VIII. 47.

The Accipiter palumbarius[1] I take to be the Sparrow-Hawk of the English and the Sperwer of the Germans, since it preys on Doves, Pigeons, and Partridges and the bigger sorts of birds.

The Fringillarius I guess to be the Hobby of the English. Now the Hobby is a very little Hawk of darker colour than the other kinds. It has upon the head two spots of deep black on a lighter ground. It catches for the most part Larks and Finches, nests on lofty trees, and is not seen in winter anywhere.

The Rubetarius I think to be that Hawk which English people name Hen-Harrier. Further it gets this name among our countrymen from butchering their fowls. It exceeds the Palumbarius in size, and is in colour ashen. It suddenly strikes birds when sitting in the fields upon the ground, as well as fowls in towns and villages. Baulked of its prey it steals off silently, nor does it ever make a second swoop. It flies along the ground the most of all.

The Subbuteo I think to be that Hawk which Englishmen call Ringtail from the ring of white that reaches round the tail. In colour it is midway from fulvous to black; it is a little smaller than the Buteo, but much more active. It catches prey in the same manner as the bird above.

Of the Alcedo.

Ἀλκυών, alcedo, in English the kynges fisher, in German eyn eissvogel.

Aristotle.

The Alcedo, not much larger than the Passer, is remarkable for being in its colour green and blue, and even slightly purple, not, that is to say, in separate parts, as if it had the colour perfectly distinct, but variably shining over every part alike of the whole body, with the wings and head. The beak is greenish, and is long and thin. The tribe of

[1] Later authors are probably more correct in applying this name to the Goshawk, which suits even Turner's account better.

que genus aquas adamat, quod duplex eſt: alterum uocale, harundinibus inſidens, alterum mutum, quod ampliore corpore eſt utrique dorſum cœruleum. Sed alcedo apud mare quoque uerſatur.

PLINIUS[1].

Ipſa auis paulò amplior paſcere[2], colore [p. 20] cyaneo, ex parte maiore, tantùm purpureis & candidis admixtis pennis, collo gracili ac procero. Alterum genus earum, magnitudine diſtinguitur, & cantu. Minores in harundinetis canunt. Halcyonem uidere rariſſimũ eſt, nec niſi Vergiliarum occaſu, & circa ſolſtitia, brumam'ue, naue aliquando circumuolata, ſtatim in latebras abeuntem. Fœtificant bruma, qui dies Halcionides uocãtur, placido mari per eos & nauigabili, Siculo maximè. In reliquis partibus eſt quidem mitius pelagus. Siculũ utiq; tractabile. Faciunt autem ſeptem ante brumam diebus nidos, & totidem ſequentibus pariũt. Nidi earum admirationem habent, pilæ figura paulũ eminente, ore perquàm anguſto, grandium [p. 21] ſpongiarum ſimilitudine, ferro intercidi non queunt, franguntur'q; ictu ualido, ut ſpuma arida maris. Nec unde confingantur inuenitur. Putant ex ſpinis aculeatis, piſcibus enim uiuunt. Subeunt & in amnes. Pariunt oua quina.

[1] *Hist. Nat.* Lib. X. cap. xxxii.

[2] Lege 'passere.'

Kingfishers, of which there are two sorts, is fond of watersides: one is a vocal bird, which sits on reeds, the other, which is of a larger size, is mute. The back is blue in both. The Kingfisher, however, also haunts the sea.

PLINY.

This bird is little bigger than the Passer, for the most part blue in colour, with the wings alone of purple mixed with white, and with a long and slender neck. Each of the two kinds may be distinguished by its size and voice. The lesser sing in reed-beds. It is very rare to see the Halcyon, and this occurs only towards the setting of the Pleiades and near the solstice or in winter-time, when, after circling round the ship awhile, it hurriedly departs again to its retreat. They breed in winter, at the season called the Halcyon days[1], wherein the sea is calm and fit for navigation, the Sicilian sea particularly so. Elsewhere indeed the ocean is less boisterous. The Sicilian is certainly gentle enough. Now these birds build their nests in the seven days before the winter solstice, and hatch out their young in the seven following. Their nests compel our wonder, of a ball-like shape, with a small jutting part and very narrow hole, like sponges of great size; they cannot be cut open with an iron tool, but may be broken by a vigorous blow, as dry sea-foam[2] may be. It is not known of what these are composed. Some think of pointed bones, since the birds live on fish. They also dive in rivers, and lay five eggs each.

[1] For the origin of this ancient tradition, the reader may be referred to any work dealing with Greek mythology.

[2] By 'dry sea-foam' Pliny probably meant masses of whelks' eggs.

Præter hæc duo ab Ariſtotele & Plinio deſcripta genera, auem noui, quæ ſi alcedonum generibus non ſit adſcribenda, ſub quo genere contineatur, prorſus neſcio. Ea ſturno paulò minor eſt, corpore toto nigro, excepto uentre albo. Caudam habet breuiuſculam, roſtrum alcedone paulò breuius. Ante uolatum, alcedonis more crebrò nutat, & in uolatu gemit: uoce alcedonẽ ita refert, ut, niſi uideas, alcedonem eſſe iurares: in ripis fluminum, non procul à mari uidi, aliâs nuſquam. piſciculis uictitat ut ſuperiora alcedonum genera. Nidum huius nunquam uidi. Morpetenſes, apud quos auem uidi, cornicem uocant aquaticam.

a uuater crauu.

DE ANATE.

Νῆττα, *anas, Anglicè a duck, Germanicè* eyn endt.

Plinius[1].

[p. 22] Anates ſolæ, quæ'que ſunt eiuſdem generis, in ſublime ſeſe protinus tollunt, atq; è ueſtigio cœlum petunt, & hoc etiam ex aqua.

DE ANSERE.

Χήν, *anſer, Anglicè a goſe, Germanicè* eyn ganß.

Ariſtoteles ſimul & Plinius duo præcipua anſerum genera faciunt: hic anſerem in maiorem & minorem, ille in domitũ et ferũ diuidens. Sed Plin. præter hæc duo anſerum præcipua genera, Penelopes[2] *et chenalopeces, ut unus textus habet, &, ut alius habet, chenalopeces, & chenerotes anſerini eſſe generis tradit. Prior lectio ſic habet,*

[1] *Hist. Nat.* Lib. x. cap. xxxviii.

[2] Judging from p. 148 of the original work the singular of this word is 'Penelops,' and it is probably by mistake that the Wigeon has been called *Mareca penelope*.

Besides the two kinds thus described by Aristotle and Pliny I know of a bird, of which, if it should not be properly ascribed to the Kingfisher tribe, I really cannot say under what head it ought to go. It is a little smaller than a Starling, with the body wholly black, except for a white belly, and it has the tail comparatively short, the beak a little shorter than the Kingfisher. Before a flight it dips repeatedly, after the manner of the Kingfisher, and cries out as it flies; it is so like the Kingfisher in voice that, if you did not see it, you would swear it was a Kingfisher. I have observed it on the banks of streams not far from the sea-side, but nowhere else. It lives on little fishes, like the aforesaid kinds of Kingfishers. I never saw its nest. The inhabitants of Morpeth, where I saw the bird, call it a water craw[1].

Of the Anas.

Νῆττα, anas, in English a duck, in German eyn endt.

Pliny.

Anates only, and birds of like kind, rise in the air at once, and make straight for the sky, and that even from the water.

Of the Anser.

Χήν, anser, in English a goose, in German eyn ganss.

Aristotle agrees with Pliny in making two chief kinds of Geese, the latter separating them into the greater and the less, the former into tame and wild. But Pliny tells us that besides these two chief kinds of Geese, there are of the Goose kind Penelopes and Chenalopeces, as one text has it, as another goes, Chenalopeces and Chenerotes. The first reading stands thus:—

[1] The bird meant is undoubtedly the Water Ousel or Dipper (*Cinclus aquaticus*), which still goes by the name of Water Craw in the north of England. It is curious that Turner should never have seen its nest when he was in Northumberland.

Anserũ generis ſunt Penelopes, & quibus lautiores epulas Britannia non nouit, chenalopeces, anſere ferè minores. *Altera ſic habet:* Anſerini generis ſunt chenalopeces & quib⁹ lautiores epulas Britannia nõ nouit, chenerotes[1].

Poſterior lectio mihi magìs approbatur, nam & nos una aue locupletat, et penelopes anatini potiùs q̃ anſerini generis eruditis eſſe uidentur. Sed quæ'nam iſtæ aues, & quibus nominibus apud noſtrates appellantur, dicere tentabo. Chenalopex, ab anſere et uulpe nomen habet, & Latinè à Gaza uulpanſer dicitur. Noſtrates hodie bergandrum nominãt, anate longior & grandior uulpanſer eſt, pectore ruffeſcente, in aquis degit, & in cuniculorũ foueis. interdum & in excelſarum rupium cauernis (unde fortè nomen ab Angloſaxonibus, noſtris patribus ſortitus eſt) nidificat. Nuſquam aliàs uulpanſerem uidi, niſi in Tamiſi fluuio. Aiunt tamen frequẽtem eſſe in inſula Tenia uocata, & illic in ſcrobibus cuniculorum nidulari. Moribus admodum uulpinis eſt. nam dum teneri adhuc pulli ſunt, ſi quis eos captare tentet, prouoluit ſeſe uulpanſer ante pedes captantis, quaſi iam capi poſſit, atq; ita allicit ad ſe capiendam hominem, eouſq; dum pulli effugiant: tum ipſe auolat & reuocat prolem. Chenerotes quæ'nam aues ſint, puto paucìſſimos hodie eſſe, qui nouerunt. Neq;

[p. 23]

a bergander.

[1] *Hist. Nat.* Lib. x. cap. xxii.

"Of the Goose kind there are Penelopes and also Chenalopeces, the latter generally smaller than a Goose; and Britain knows no richer feast than these."

The second runs:—

"Of the Goose kind are Chenalopeces and Chenerotes, Britain knows no richer feast than these."

To me the latter reading most approves itself, for it both makes us richer by one bird, and the Penelopes seem to our learned men to be of the Duck tribe rather than of the Goose. But I will try to say what these birds are and by what names they go among our countrymen. The Chenalopex[1] takes its name from the Goose and the Fox, while it is called by Gaza Vulpanser in Latin, though our people nowadays name it Bergander[2]. It is longer than a Duck and bigger, with a ruddy breast. It lives upon the waters and in coneys' burrows. At times it even nests in holes of lofty rocks (whence possibly the name was first allotted to it by our ancestors the Anglo-Saxons). I have nowhere else seen the Vulpanser save upon the river Thames. Nevertheless they say that it is plentiful upon the isle which is called Tenia[3], and that it breeds in coneys' burrows there. In habits it is very like a Fox, for, while the young are still of tender age, should any one attempt to capture them, the old Vulpanser rolls upon the ground before his very feet[4], as if she could be taken there and then, and thus allures the man to follow her, until the young are able to escape; then she flies off and summons back her brood. I think that there are very few men now who know what sort of birds the

[1] Turner's bird was undoubtedly the Sheld-Drake (*Tadorna cornuta*), notwithstanding the fact that the name *Chenalopex* has been conferred on the so-called 'Fox-Goose' of Africa.

[2] The Sheld-Drake is still the Bargander or Bergander of some districts of England; possibly the correct spelling should be Burgander, i.e. Burrow Duck. The word seems to have nothing to do with Berg = a mountain.

[3] Possibly St Mary's, or even Coquet Island.

[4] The Sheld-Drake does not usually behave thus.

ego, licet Britannus, chenerotes noſtros ſatis noui: nā præter duo Ariſtot. genera, anſerū adhuc duo genera noui in Britānia, ad quorū neutrum ſi chenerotes pertineāt, chenerotes mihi penitus ignotos eſſe ingenuè fatebor. Prior anſer à noſtris hodie brāta & bernicla uocatur, & fero anſere minor eſt, pectore aliquò uſq; [p. 24] *nigro. Cætero cinerio, anſerum ferorū more uolat, ſtrepit, paludes frequētat, & ſegetē depopulatur. Caro huius paulò inſuauior eſt, & diuitibus minùs appetita. Nidum berniclæ, aut ouum nemo uidit: nec mirum, quum ſine parentis opera berniclæ ad hunc modū ſpontaneam habeāt generationē. Quum ad certum tempus, malus nauis in mari cōputruit, aut tabulæ, aut antennæ abiegnæ, inde in principio ueluti fungi erumpūt: in quibus temporis progreſſu, manifeſtas auiū figuras cernere licebit, deinde pluma ueſtitas, poſtremò uiuas & uolantes. Hoc, ne cui fabuloſum eſſe uideatur, præter cōmune omniū gentiū littoraliū Angliæ, Hiberniæ & Scotiæ, teſtimoniū Gyraldus ille præclarus hiſtoriographus qui multò fœlicius q̃ pro ſuo tempore Hiberniæ hiſtoriam conſcripſit, nō aliam eſſe berniclarū generationē teſtatur. Sed, quum uulgo non ſatis tutū uideretur fidere, et Gyraldo ob rei raritatem non ſatis crederem, dum hæc, quæ nunc ſcribo, meditarer, uirum quendam, cuius mihi perſpectiſſima integritas fidem merebatur, profeſſione Theologum, natione Hibernum, nomine Octauianū, conſului num Gyraldum hac in re fide dignum cenſeret? qui per ipſum iurans, quod profitebatur euangelium, reſpondit, ueriſſimum eſſe, quod de generatione huius auis Gy-* [p. 25] *raldus tradidit, ſeq; rudes adhuc aues oculis uidiſſe, & manibus contrectaſſe: breuiq; ſi Londini menſem unum aut alterum manerem, aliquot rudes auiculas mihi aduectas curaturū. Iſta berniclæ generatio nō uſq; adeo*

Chenerotes are. And, though I am a Briton, I am not quite sure about our Chenerotes; for as yet, apart from the two kinds that Aristotle gives, I know two sorts of Geese in Britain and will frankly own that, if the Chenerotes are not to belong to either of them, they are quite unknown to me. The first Goose by our people nowadays is called the Brant and Bernicle, and is a smaller bird than the Wild Goose, with the breast partly black. The rest is ashen grey. It flies, gabbles, haunts swamps, and devastates green crops, like the Wild Goose. Its flesh is somewhat strong, and is the less sought after by the rich. No one has seen the Bernicle's nest or egg, nor is this wonderful, since Bernicles without a parent's aid are said to have spontaneous generation in this way: When after a certain time the firwood masts or planks or yard-arms of a ship have rotted on the sea, then fungi, as it were, break out upon them first, in which in course of time one may discern evident forms of birds, which afterwards are clothed with feathers, and at last become alive and fly. Now lest this should seem fabulous to anyone, besides the common evidence of all the long-shore men of England, Ireland, and Scotland, that renowned historian Gyraldus[1], who composed a history of Ireland in much more happy style than could have been expected in his time, bears witness that the generation of the Bernicles is none other than this. But inasmuch as it seemed hardly safe to trust the vulgar and by reason of the rarity of the thing I did not quite credit Gyraldus, while I thought on this, of which I now am writing, I took counsel of a certain man, whose upright conduct, often proved by me, had justified my trust, a theologian by profession and an Irishman by birth, Octavian by name, whether he thought Gyraldus worthy of belief in this affair. Who, taking oath upon the very Gospel which he taught, answered that what Gyraldus had reported of the generation of this bird was absolutely true, and that with his own eyes he had beholden young, as yet but rudely formed, and also handled them, and, if I were to stay in London for a month or two, that he would take care that some growing chicks should be brought in to me. This curious generation of the Bernicle will not appear so very

[1] Giraldus Cambrensis, *Topographia Hibernica* Distinctio I. cap. xv.

prodigiosa illis uidebitur, qui quod Aristoteles de uolucre ephemero scripsit, legerint.

De ephimero autem Aristoteles[1] *libro quinto de historia animalium ita scribit.* Hyppanis fluuius apud Cymerium Bosphorum sub solstitio, defert ueluti folliculos acinis maiores, quibus quadrupedes uolucres erumpunt: quod genus animalis in postmeridianum[2] usque diei tempus uiuit & uolat: mox descendente sole, macrescit & languet[3]: deinde occidente, moritur, uita non ultra unum diem protracta: unde ephemerum, id est, diarium[4] appellatum est. Hęc Aristotel.

Quæ si uera sunt, & tāto philosopho digna, superioris auis generationi non parum fidei adstruent.

a solend gose.

Alter anser, de quo promisi me dicturū, marina [p. 26] *auis est, ex uenatu piscium uictitans, magnitudine superiore ansere paulò minor: anserem tamen uoce & forma per omnia refert, nidulatur in mari Scotico, in rupibus excelsis, insulæ Bassi, per antiphrasim, opinor, dictæ: nec alias uspiam in tota Britannia. Hic tanto amore suos pullos prosequitur, ut cum pueris per funes in corbibus ad auferēdos eos demissis, acerrimè non sine uitæ periculo confligetur. Nec silentio prætereundum est, ex adipe huius anseris (est enim insigniter adiposus) unguentum à Scotis ad multos morbos utilissimum fieri, quod cum commageno à Plinio*[5] *celebrato, meritò bonitate & remediorum numero potest certare. Iam quū anserum genera, licet diligentissimè inquirēs, apud Britannos plura inuenire non possim, chenerotes*

[1] Bk V. 107.

[2] μέχρι δείλης.

[3] These two words are not in the original Greek.

[4] This explanation is not given by Aristotle. We have here an instance of the insertions common in old authors, which will not be noticed hereafter in each case, as being too numerous. Another instance is found with regard to 'Albicilla' (p. 30).

[5] *Hist. Nat.* Lib. X. cap. xxii.

marvellous to those who may have read what Aristotle wrote about the flying creature called Ephemerus. Now Aristotle writes thus of the Ephemerus in the fifth book of his History of Animals:—

"The river Hyppanis[1], near the Cymerian Bosphorus[2] when the solstice is nigh, brings down small pouches, as it were, each larger than a grape, from which four-footed flying creatures burst; a sort of animal which lives and flies until the afternoon of the same day, but presently at the sun's going down withers and languishes, and finally, at the sun's setting, dies, lasting no longer than a single day, whence it is called Ephemerus, that is, the creature of a day." Thus Aristotle writes.

Now if these things are true, and worthy of the great philosopher, they will impart no little credibility as to the generation of the aforesaid bird.

The second Goose, of which I promised I would speak, is a sea-bird, which lives by hunting fishes, somewhat less in size than the Goose given above; and yet in voice and aspect it recalls the Goose in every way; it nests within the Scottish sea, upon the lofty cliffs of the Bass Isle—so called, as I opine, by an antiphrasis[3]—and nowhere else in all Britain. This bird looks to its young with so much loving care, that it will fight most gallantly with lads that are let down in baskets by a rope to carry them away, not without danger of its life. Nor must we fail to mention that a salve, most valuable for many a disease, is made by Scots from the fat of this Goose (for it is wonderfully full of fat) which may deservedly rival the Commagenum vaunted much by Pliny, in its virtue and the number of its cures.

Now since, though searching with the greatest care, I cannot find any more kinds of Geese among Britons,

[1] Now the Bog.
[2] Between the Sea of Azov and the Black Sea.
[3] As if the derivation was from the French *bas*=low.

(qui ab amore mihi nomen habere uidentur), aut berniclæ aut Baſſani anſeres ſunt, aut mihi prorſus ignoti.

DE AQVILA.

ἀετὸς, *aquila, Anglicè an egle, Germanicè* ein ärn, oder ein adler.

ARISTOTELES[1].

Aquilarum plura ſunt genera. Vnum, quod [p. 27] pigargus ab albicante cauda dicitur, ac ſi albicillam nomines. gaudet hęc planis, & lucis et oppidis. Hinnularia[2] à nonnullis uocata cognomine eſt. montes etiam, ſyluamq́, ſuis freta uiribus, petit. reliqua genera rarò plana & lucos adeunt.

Pygargus, quum ſit primum aquilarum genus, Germanorum literatores turpiter errant, qui pygargum ſuum trappum faciunt, qui apud Ariſtotelem tetrix, & Plinio tetrao eſt, ut poſtea docebo. Pygargus Anglorum lingua, niſi fallar, erna uocatur. an erne.

DE PLANGA AUT CLANGA EX ARISTOTELE[3].

Alterum genus magnitudine ſecundnm & uiribus, clanga[4] aut planga nomine, ſaltus & conualles, & lacus incolere ſolitum, cognomine anataria[5], & morphna, à macula pennæ, quaſi neuiã[6] dixeris, cuius etiam meminit Homer. in exitu Priami[7].

[p. 28]

PLINIUS[8] DE MORPHNA SIVE PLANGA.

Morphnos, quam Homerus & percnon uocat, aliqui & plancum & anatariã, ſecũda magnitu-

[1] *Hist. An.* Bk IX. 111.
[2] νεβροφόνον = fawn-slayer.
[3] *Hist. An.* Bk IX. 112.
[4] For πλάγγος some texts have πλάνος. The word 'clanga' does not seem to be represented in the Greek.
[5] νηττοφόνος = duck-slayer.
[6] This explanation is not in Aristotle.
[7] *Iliad*, Bk XXIV. l. 316.
[8] *Hist. Nat.* Lib. X. cap. iii.

the Chenerotes (which seem to me to get their name from "love"[1]) are either Bernicles, or the Geese of the Bass, or are decidedly unknown to me.

Of the Aquila.

ἀετὸς, aquila, in English an eagle, in German ein ärn, or ein adler.

Aristotle.

Of Aquilæ there are several kinds. One which is called Pygargus from its whitish tail, as though you were to name it Albicilla, loves plains, groves, and towns. For by-name it is called by certain Hinnularia. It even seeks the mountains and the wood, relying on its might. The other kinds seldom approach the plains and groves.

Now, seeing that Pygargus is the first kind of the Aquilæ, the German scribblers err disgracefully, who reckon it their Trapp, which is the Tetrix in the works of Aristotle and the Tetrao of Pliny, as I shall shew afterwards. Pygargus, if I err not, in the English tongue is called an Erne.

Of the Planga or Clanga, from Aristotle.

Another kind, second in size and strength, by name Clanga or Planga, generally haunts glades and valleys and lakes. It has the by-name Anataria, and Morphna from the marking on the wing, as though you should say spotted. Of this Homer makes mention in the scene of Priam's death.

Pliny on the Morphna or Planga.

Morphnos, which Homer also calls Percnos, some name Plancus and Anataria, second in size and

[1] A very doubtful derivation.

dine & ui, huicq; uita circa lacus. Ista circa stagna aquaticas aues appetit mergẽtes se subinde, donec sopitas lassatasq; rapiat. Spectanda dimicatio, aue ad perfugia littorum tendente, maximè si condensa harundo sit: aquila †inde ictu abigente ala, & cùm appetit in lacus cadente, umbramq; suam nanti sub aqua à littore ostendente: rursus aue in diuersa, & ubi minimè se credat expectari, emergente. Hæc causa est gregatim auibus natandi, quia plures simul non infestantur: respersu pinnarũ †hostẽ obcæcantes. Sæpè & aquilæ ipsæ non tollerantes pondus apprehensum, unà merguntur. Hæc Plinius.

† in deiectu.

† hostem aliâs abũdat.

[p. 29]

Omnia, quæ Aristoteles & Plinius percno hactenus tribuerunt, Anglorum balbushardo conueniunt, si solam magnitudinem exceperis, quæ si alia adfuerint, hic fortassis non oberit. est autem illa, quam anatariam esse conijcio, auis buteone maior & longior, neuo albo in capite, colore fusco proximo, ad ripas fluminum, stagnorum et paludium semper degens, uiuit ex uenatu anatum et gallinarum nigrarum, quas Angli coutas nominãt. Venationem hanc, cuius meminit Plinius, inter aquilam istam (si aquila dicenda sit) & aues aquaticas, non solùm ego sæpissimè uidi, sed infiniti apud Anglos quotidie uident. Si qua terræ portiuncula super aquas inter arundineta emineat, in hac solet nidum facere, ut quoniam uolatu non admodum ualet, à præda non procul absit. Aues subitò adoritur, & sic capit. Cuniculos ista interdum etiam dilaniat. Nunc an ista anataria sit nec ne, doctis uiris iudicandum propono.

strength; it passes its life round lakes. By pools it chases water-birds, which dive from time to time, until it catches them sleepy and weary. The contest is a sight to see, the quarry seeking refuge on the shore, chiefly where reeds are thick, and thence the Aquila drives it away with a stroke of the wing and plunges in the lake as it swoops from above, shewing its shadow to the bird as it swims under water from the shore. Again the latter tries a different place and comes up where it thinks that it will least be marked. This is the cause of birds swimming in flocks, for they are not molested when in companies, and blind their enemy by splashing with their wings. The Aquilæ themselves, moreover, often are immersed, not being able to support the weight that they have clutched. Thus Pliny.

All things that Aristotle and Pliny have attributed to the bird Percnos so far well agree with the Balbushard of the English[1], if one may except its size alone, and if the rest be present, that perhaps should not stand in the way. Now the bird which I apprehend to be the Anataria, being bigger and longer than the Buteo, with a white patch upon the head, and nearly fuscous in colour, always haunts the banks of rivers, pools, and swamps; it lives by hunting Ducks and those black fowls which Englishmen call Couts. The conflict of which Pliny makes mention above between this Eagle (if it should be called an Eagle) and the water-birds I have seen often, and not I alone, but countless Englishmen witness it daily. If anywhere a little space of ground rises among the reed-beds, there the bird is wont to make a nest, that, since in power of flight it is not very strong, it may not be far distant from its prey. It suddenly attacks birds, and thus takes them. It also sometimes butchers coneys. Now whether this may be the Anataria or not I put it to the learned to decide.

[1] The Bald-Buzzard or Marsh-Harrier (*Circus æruginosus*).

De tertio genere ex Aristotele[1].

Tertium genus colore nigricãs, unde nomen [p. 30] accepit, ut pulla & fuluia[2] uocetur, magnitudine minima, ſed uiribus omnium prę ſtantiſſima. Hæc colit montes & ſyluas & leporaria cognominatur.

Plinius[3].

Melænaëtos. Melænaëtos à Græcis dicta, eademq; ualeria, minima magnitudine, uiribus præcipua, colore nigricans: ſola aquilarum fœtus ſuos alit, cæteræ fugant: ſola ſine clangore, ſine murmuratione.

De quarto genere ex Aristot.[4]

percnopterus. Quartum genus percnopterus ab alarum notis, capite albicante, corpore minore, quàm cæteræ adhuc dictæ, hæc eſt. Sed breuioribus alis, cauda longiore, uulturis ſpeciem hęc refert. Subaquila[5], & aquila montana cognominatur. Incolit [p. 31] lucos, degener, nec uicijs cæterarum caret, & bonorum, quæ illæ obtinent, expers eſt: quippe quæ à coruo, cæterisq; id genus auibus uerberetur, fugetur, capiatur. Grauis enim eſt, uictu iners: examinata[6] fert corpora: famelica ſemper eſt, et querula, clamitat, & clangit.

DE HALIÆETO.

Haliæetus Græcè & Latinè, Anglicè an oſprey, Germanicè eyn viſhärn.

[1] *Hist. An.* Bk IX. 113.

[2] Aristotle has merely: καλεῖται δὲ μελανάετος καὶ λαγωφόνος.

[3] *Hist. Nat.* Lib. X. cap. iii.

[4] *Hist. An.* Bk IX. 114; a very free version.

[5] Some texts read γυπαίετος for ὑπαιετός.

[6] Lege 'exanimata.' Aristotle has τὰ τεθνεῶτα φέρων.

Of the third kind from Aristotle.

The third kind in colour is blackish, whence it has received its name, so that the bird is called Pulla and Fulvia, in size the least of all and yet chiefest in strength. It haunts mountains and woods, and is called also Leporaria.

Pliny.

The bird called Melænaetos among the Greeks, which is the same as the Valeria, is very small in size, but chief in strength, in colour blackish: of the Aquilæ this kind alone fosters its young, the others drive them off: it is the only one without a scream, without a softer note.

Of the fourth kind from Aristotle.

The fourth kind, called Percnopterus, from having spots upon the wings, is whitish on the head; it has a smaller body than the other sorts spoken of hitherto. But with its shorter wings and longer tail it has the aspect of a Vulture. It is called besides Subaquila and Mountain Aquila. It dwells in woodlands, an ignoble bird, not lacking the bad qualities of others, but void of the good that they possess. For it is beaten, put to flight, and caught by the Raven and by other birds like that. Further it is unwieldy, sluggish to get food, and carries off dead bodies; it is always ravenous and querulous; it cries continually and screams.

Of the Haliæetus.

Haliæetus in Greek and Latin, in English an Osprey, in German eyn vishärn.

Plinius[1].

Supereſt Haliæetus, clariſſima oculorum acie, librans ex alto ſeſe uiſoq; in mari piſce, præceps in eũ ruens, & diſcuſſis pectore aquis rapiens.

Aristoteles[2].

Haliæetos, hoc eſt, marina aquila, ceruice [p. 32] magna & craſſa, alis curuantibus, & cauda lata eſt. Moratur hæc in littoribus & oris. Accidit huic ſæpius, ut quum ferre, quod ceperit, nequeat, in gurgitem demergatur.

Haliæetos apud Anglos hodie notior eſt, quàm multi uelint, qui in uiuarijs piſces alunt: nam piſces omnes breui tempore aufert. Piſcatores noſtrates eſcis fallendis piſcibus deſtinatis, haliæeti adipem illinunt, aut immiſcent, putantes hoc argumento eſcam efficaciorẽ futuram, quòd haliæeto ſeſe in aëre librãte, piſces quotquot ſubſunt (natura aquilæ ad hoc cogente, ut creditur) ſeſe reſupinẽt, & uẽtres albicantes, ut quem liberet, eligeret, exhibeãt.

De aquila vera ex Aristotele[3].

Angli. a right egle German. eyn edel arn.

Sextum genus gneſium, id eſt, uerũ germanumq; appellant. Vnũ hoc ex omni aquilarum genere, ueri incorruptiq; ortus creditur. Maxima omnium aquilarum hæc eſt, maior etiam [p. 33] quàm oſſifraga[4]: ſed cæteras aquilas uel ſeſquialtera portione excedit, colore ruffa eſt, conſpectu rara.

DE ARDEA.

ἐρωδιός, *ardea, Anglicè an heron. Germanicè* eyn reyger.

[1] *Hist. Nat.* Lib. x. cap. iii.

[2] *Hist. An.* Bk ix. 115, a free version.

[3] *Hist. An.* Bk ix. 116.

[4] Gaza translates φήνη by ossifraga, but it is very doubtful what bird the ossifraga really was. Possibly it should be identified with the Lämmergeier (cf. Prof. Newton, *Dict. Birds*, p. 660).

PLINY.

The Haliæetus remains, with eyesight of the keenest, poising itself aloft when it spies fishes in the sea below, then dashing headlong on them and securing them, the waters being parted by its breast.

ARISTOTLE.

The Haliæetos, that is to say Sea Eagle, has the neck both big and thick, bowed wings, and a broad tail. It bides upon the sea-coast and the shores. It often happens, when it cannot lift what it has taken, that it is submerged beneath the tide.

The Osprey is a bird much better known to-day to Englishmen than many who keep fish in stews would wish; for within a short time it bears off every fish. Our anglers smear or mix their bait with Osprey's fat, arguing that thus the bait will prove more efficacious from the fact that, when the Osprey hovers in the air, whatever fishes be below turn up and shew their whitish bellies (as it is believed, the nature of the Aquila compelling them to this), that it may choose that one which it prefers.

OF THE TRUE AQUILA FROM ARISTOTLE.

The sixth kind men call Genuine, or true and thoroughbred. Of all the various kinds of Aquilæ this is the only one that is believed to be of true and unstained origin. This is the largest of all Aquilæ and bigger even than the Ossifrage, for it surpasses by one half as much the other Aquilæ; in colour it is reddish brown, but it is rarely seen.

OF THE ARDEA.

ἐρωδιός, ardea, in English a heron, in German eyn reyger.

ARISTOTELES[1].

Ardearum tria ſunt genera, pella, alba, ſtellaris, piger cognomine. Pellæ coitus difficilis eſt: uociferatur enim, & ſanguinem ex oculis (ut aiunt) emittit cùm coit; parit etiam ægrè ſummoq; cum dolore. Pella ſagax[2] & cœnæ gerula eſt, & operoſa[3]. Agere interdiu ſolet: colore tamen & prauo & aluo humida. Reliquarum duarum, alba colore eſt pulchro, & coit, & nidulatur & parit probè, paſcitur paludibus, lacu, campis & pratis. Sed ſtellaris piger cognominata, (in fabula est, ut olim è ſeruo in auem tranſierit) atque, ut cognomẽ ſonat, iners ocioſaq; eſt. Phoici appellatæ[4], peculiare præ cæteris eſt, ut oculos potiſſimùm appetat[5]. Petit lacus & fluuios ardea[6] & albardeola, quæ magnitudine minor eſt, roſtro lato, porrectoq;.

Pella. Stellaris. [p. 34]

Pella apud Anglos in excelſis arboribus, nõ procul à ripis fluminum creſcentibus nidum facit. Superior pars corporis cyanea eſt, inferior autem nõnihil candicat, uentris excremẽtis liquidioribus inuadentes ſe ſubitò aquilas, aut accipitres abigit, & ſe ita defendit. Vidi & huius generis, licet raras, albas, quæ neque corporis magnitudine, neque figura, ſed ſolo colore, à ſuperiore diſtulerunt. Viſa eſt etiam alba cum cyanea apud Anglos nidulari, & prolem gignere. Quare eiuſdem eſſe ſpeciei, ſatis conſtat. Albardeolam, quæ Græcè λευκερωδιός *dicitur, ſemel tantùm in Italia uidi, pella multò minor eſt, & hominis conſpectũ nõ perinde atq; cærulea fugit. Hãc ſi nõ uidiſſem, Anglorũ shoue-*

The blue heron.

a cryel heron or a duuarf heron. a myre dromble.

[1] *Hist. An.* Bk IX. 19.

[2] *Hist. An.* Bk IX. 93.

[3] Aristotle's word is ἔπαγρος, which Sundevall renders by the Swedish equivalent of 'forages round the fields.'

[4] *Hist. An.* Bk IX. 94.

[5] This seems to mean that the φῶϋξ *eats* other creatures' eyes, for Aristotle says: μάλιστα γαρ ἐστιν ὀφθαλμοβόρος τῶν ὀρνίθων.

[6] *Hist. An.* Bk VIII. 46.

ARISTOTLE.

Of Ardeæ there are three kinds, Pella, Alba, and Stellaris, but the last has the by-name of Piger. The coupling of Pella is difficult, for it screams while it couples and (they say) emits blood from its eyes: it also brings forth painfully and with extreme distress. The Pella is sagacious, quick at getting food[1], and always busy. It is wont to be astir by day; yet it is mean in colour, with the belly wet. Of the remaining two the Alba, fair of colour, couples, nests and brings forth well; it feeds in marshes, on a lake, in fields and meadow-ground. But the Stellaris, by-named Lazy (in the fable it is said of old to have been changed from a slave to a bird), as its by-name imports, is slow and indolent. The bird called Phoix has beyond all others this peculiarity that it especially attacks the eyes. The Ardea and the Albardeola, which is of smaller size and has a broad and elongated bill, seek lakes and rivers.

The Pella builds its nest in England on the lofty trees that grow not far from the banks of streams. The upper part of the body is blue, the lower is, however, somewhat white. It routs Eagles or Hawks, if they attack it suddenly, by very liquid mutings of the belly, and thereby defends itself. Of this kind I have seen some white, though they are rare, which differed from the aforesaid neither in their size nor shape of body, but in colour only. Furthermore the white has been observed in England to nest with the blue, and to bear offspring. Wherefore it is clear that they are of one species. I have only once seen—and that was in Italy—the Albardeola, which is called λευκερωδιός in Greek; it is much smaller than the Pella and by no means shuns the sight of man so much as does the blue. Had I not seen it, I should have declared the Albardeola to be the English

[1] That is, for its young.

lardam albardeolā eſſe iudicaſſem. Stellaris eſt, quā [p. 35] *Angli buttourum, aut bittourum, & Germani pittourum & roſdommum nominant: nam auis eſt toto corporis habitu ardeis reliquis ſimilis, ex piſcium uenatu ad ripas paludium & amnium uiuens, pigerrima & ſtolidiſſima, ut quæ in retia ab equo facticio agi poteſt facilimè. Colore eſt ferè, quantum memini, phaſiani, roſtro limo indito, aſininos ronchos uoce refert: oculos hominum auidiſſimè omnium auium appetit. Quare ſi quid impediat, quô minùs ſtellaris eſſe poſſit, (quod mihi nondum cernere datum eſt) phoica eſſe oportebit, quam Ariſtoteles oculos maximè appetere teſtatur, quanquam & cæteræ ardeæ idem facere ſæpè uiſæ ſunt.*

a buttour
ein pittour.
ein roſdō.

Phoix.

DE AVRIVITTE.

Χρυσομίτρις, *nō ut quidam codices habent,* ῥυσομήτρης, *auriuittis, Anglicè a gold finche, Germanicè* eyn diſtelfinck, oder eyn ſtigelitz.

Auriuittis.

Auriuittis una eſt ex auiculis, quæ carduorum ſemine uictitant[1], *& uermes etiam oblatos, non attingunt. Alij goldfincam, aut diſteluincam, ſpinum, alij carduelem eſſe uolunt. Sed ſi quis, ex ſpiniuoris præter hanc aliam aurea uitta redimitam oſtēderit, cui magis auriuittis nomen competat, quàm huic, opinionem meam facilè patiar explodi, alioqui non uideo, quin digna ſit, quæ probetur.*

[p. 36]

DE ATTAGENE.

Ἀτταγὴν, ἀττάγας, *attagen, attagena.*

Attagen, ut ſcribit Ariſtoteles, gallinagini ſimilis eſt colore. Attagenam uarijs diſtinctam eſſe maculis, Ariſtophanes[2] *his uerſibus teſtatur:*

Si quis ex uobis erit fugitiuus atq; uſtus notis,
Attagen ſanè apud nos uarius appellabitur.

[1] Aristotle's groups of Birds are as follows: (1) γαμψώνυχες (crooked-clawed); (2) σκωληκοφάγα (worm-eating); (3) ἀκανθοφάγα (eating thistle seeds); (4) σκιποφάγα (? grub-eating); (5) περιστεροειδῆ (dove-like); (6) σχιζόποδα (cleft-footed); (7) στεγανόποδα (wholly webbed); (8) βαρέα (heavy, i.e. ground kinds). A few Birds, however, can hardly be placed under any of these.

[2] *Aves*, ll. 761—762.

Shovelard[1]. Stellaris is that kind which Englishmen denominate buttour or bittour, and the Germans call pittour or rosdom. Now it is a bird like other Herons in its state of body generally, living by hunting fishes on the banks of swamps and rivers, very sluggish and most stupid, so that it can very easily be driven into nets by the use of a stalking horse. So far as I remember, it is nearly of the colour of a Pheasant, and the beak is smeared with mud; it utters brayings like those of an ass. Of all birds it aims at mens' eyes most readily. Wherefore if anything hinders this kind from being the Stellaris (which is not yet given to me to see) it ought to be the Phoix, inasmuch as Aristotle testifies that it aims chiefly at the eyes, though other Ardeæ also often seem to do the same.

Of the Aurivittis.

Χρυσομίτρις (not as some texts have it ῥυσομήτρης), aurivittis, in English a gold finche, in German eyn distelfinck or eyn stigelitz.

The Aurivittis is one of the small birds that feed on seeds of thistles, and do not touch worms even when offered to them. Some will have it that the Goldfinc or the Distelvinc is but the Spinus[2], some the Carduelis. But if anyone can shew another of the thistle-eating birds save this, girt with a golden band, to which the name of Aurivittis is more fitting than to this, I gladly will allow my opinion to be ignored, but otherwise, I do not see why it should not be worthy of approval.

Of the Attagen.

Ἀτταγὴν, ἀττάγας, attagen, attagena.

The Attagen, as Aristotle writes, is like the Gallinago in colour. And Aristophanes bears witness in these lines that the Attagena is marked with varied spots:—

"If any of you be a runaway and branded with the marks, he shall assuredly be called with us the spotted Attagen."

[1] That is, the Spoonbill of modern books (*Platalea leucorodia*), while the buttour is of course the Bittern (*Botaurus stellaris*)

[2] Turner himself considered Spinus to be the Greenfinch (cf. p. 85 of the original).

PLINIUS[1] DE ATTAGENE.

Attagen maximè Ionius celebratur, uocalis aliàs, captus uerò obmutefcit, quondam exiftimatus inter raras aues. Iam & in Gallia Hifpaniaq; capitur, & per alpes.

PETRUS GYLLIUS[2].

Attagen, eft perdice paulò maior, uerficoloribus picta plumis in dorfo, & color ruffus eft, uefcitur grano, breuibus eft alis, & puluerator eft.

Falluntur igitur Britannici ludimagiftri, qui fuũ Wodcoccum attagenem faciunt, qui folis uefcitur uermibus, & grana nunquam attingit. An attagenes apud [p. 37] *Anglos inueniantur nec'ne, multùm fanè ambigo. nam qui attagenem defcribunt, marem à fœmina non feparant. unde colligo eundem fuiffe colorem, & eandem figuram maris & fœminæ. Cæterum in hoc auium genere, quod apud nos ad attagenis formam proximè accedit, mas à fœmina ita differt ut duorum generũ iftiufmodi rerũ inperito uideri poffint. Vtranque tamen auem defcribam.*

Mas gallo domeftico paulò minor, totus niger eft, excepta ea parte caudæ, quæ podicem tegit, ea enim alba eft. Cæterũ nigredo huius nonnihil fplendefcit, ad eum ferè modum, quo columborum nigrorum torques circa colla fplendefcunt. ad uiriditatem igitur proximè accedit. in capite rubrum quendam habet, fed carneum cirrũ, & circa genas duos habet ueluti lobos rubros, & eos carneos. Fœmina tota maculis diftincta eft, & à perdice, nifi maior effet, & ruffa magis, ægrè dignofci

[1] *Hist. Nat.* Lib. X. cap. xlviii.

[2] Petrus Gyllius was the author of the work *De vi et natura Animalium etc.* Lugd. Bat. 1533.

Pliny of the Attagen.

The Attagen is most renowned as an Ionian bird; it usually is noisy, in captivity however it is dumb. In former times it was considered rare, but now it is taken in Gaul, in Spain, and on the Alps.

Petrus Gyllius.

The Attagen is rather larger than the Perdix, and is marked with particoloured feathers on the back, in colour it is reddish, and it feeds on grain. It has short wings, and rolls itself in dust.

Accordingly our British schoolmasters are wrong who make their Woodcock the Attagen, which lives only on worms and never touches grain. Indeed I seriously doubt whether Attagenes be found in England or not, for those who give descriptions of the Attagen, do not distinguish the male from the female, whence I infer that they have the same colours and are like in form. But in the kind of bird which with us comes the nearest to the Attagen in form, the male differs so greatly from the female that they might appear to be of separate kinds to the man inexperienced in things like this. Nevertheless I will describe each bird.

The male[1] is somewhat less than a domestic cock and is entirely black, save that part of the tail which overlies the vent, for that is white. Moreover the black colour of the bird is somewhat glossy, very nearly as the collar round the neck of our black pigeons is. So it approaches very near to green. Upon its head it has a red but fleshy sort of comb[2], and round its cheeks two red lobes as it were and those fleshy. The hen is wholly marked with spots, and, were she not a bigger bird and more rufous, could scarcely be distinguished from a Partridge. Both frequent

[1] Turner here undoubtedly refers to the Blackcock (*Tetrao tetrix*).

[2] The Blackcock has two erectile patches of red skin over the eyes, which in the breeding season even reach above the top of the head; and the word 'cirrum' must be taken to mean such a patch here.

poſſit. In deſertis locis & planis, erica potiſſimum conſitis, ambo degunt. grano ueſcuntur, et ſummis ericæ germinibus. Breues habens alas, & breues faciunt uolatus. Hæc auis, ſi attagen nō ſit, gallina uidetur eſſe Varronis ruſtica. Eraſmus in Adagio, Attagenæ nouilunium, attagenam auem paluſtrem facit, & uarijs maculis diſtinctam. quod ſi ſatis exploratum [p. 38] *mihi eſſet, Anglorum goduuittam, ſiue fedoam, attagenam eſſe, indubitanter auderem adfirmare. Eſt autem ipſa gallinagini ita ſimilis, ut niſi paulò maior eſſet, & pectoris color magìs ad cinereū uergeret, altera ab altera difficulter poſſit diſtingui. uiuit in locis paluſtribus, et ad ripas fluminū. roſtrum habet longum, ſed capta triticum non ſecus atque columbi, comedit. triplo pluris quàm gallinago apud nos uenditur, tantopere eius caro magnatum palatis arridet. harum ſi neutra ſit attagena, attagenam nuſquam uidi.*

DE ATRICAPILLA.

Μελανκόρυφος, *atricapilla, Germa. ut creditur,* eyn graſmuklen.

Aristoteles[1].

Atricapillam etiam plurima edere aliqui referunt, ſed poſt Africam ſtrutionem. Iam uel decem & ſeptē oua atricapillæ reperta ſunt. ſed plura etiam quàm uiginti parit, & numero impari ſemper, ut narrant. Nidificat ea quoq; in arboribus, & uermiculis alitur. Proprium huius [p. 39] & luſciniæ præter cæteras aues, ut linguæ ſummæ acumine careant. Ficedulæ & atricapillæ[2] uicibus commutantur. Fit enim ineunte autumno ficedula, ab autumno protinus atricapilla,

[1] *Hist. An.* Bk IX. 88.
[2] *Hist. An.* Bk IX. 256—257.

waste open places, and especially those covered with heather. They feed on grain and on the topmost buds of heather. They have short wings and take short flights. This bird, if it be not the Attagen, appears to be Varro's Gallina rustica. Erasmus in his proverb of "the Attagena's new-moon" makes his Attagena a marsh-bird, marked with varied spots. If this approved itself sufficiently to me I confidently would venture to affirm that the Attagena was what the English call the Godwitt or Fedoa[1]. Furthermore the bird is so much like the Woodcock, that, if it were not a little larger, and did not the breast verge upon ash-colour, the one of them could hardly be distinguished from the other. It is found in marshy places and on river banks. The beak is long; but in captivity it feeds on wheat, just as our Pigeons do. With us it sells for thrice as much again as any Woodcock, so much does its flesh tickle the palates of our magnates. Of these two, if neither be the Attagena, then I have nowhere seen the Attagena.

Of the Atricapilla.

Μελαγκόρυφος, atricapilla, in German, as is supposed, eyn grasmuklen.

Aristotle.

The Atricapilla, as some report, lays the most eggs of all, next to the Struthio of Africa. No fewer than seventeen eggs of the Atricapilla have been found, but it lays even more than twenty and, as some narrate, in number always odd. It also nests in trees and feeds upon small worms. It is peculiar to this and the Luscinia beyond all other birds that they have no point to the tip of the tongue. Ficedulæ and Atricapillæ change in turn. For when autumn sets in the bird becomes a Ficedula, from autumn onwards it becomes an Atricapilla, nor is

[1] See Prof. Newton's *Dict. Birds*, p. 248.

nec inter eas diſcrimen aliquod, niſi coloris & uocis eſt. Auem eſſe eandem cōſtat, quia dum immutaretur hoc genus, utrūque conſpectum eſt, nondum abſolutè mutatum, nec alterutrum adhuc proprium ullum habens appellationis[1]. Hæc Aristoteles.

Atricapillam in Anglia nunquam uidi, neq; ſæpius in uita quàm ſemel, idq; in Italia in domo Ducis Ferrarienſis. Eamq; mihi uir utriuſq; linguæ nō uulgariter doctus, D. Franciſcus, duci à ſacris concionibus exhibuit. Anglorū lingettæ, & Germanorum graſmuſcho, quod ad corporis magnitudinem attinet, ſimilis erat: ſed atrum habebat caput, & reliquum corporis colorem magis ad cinerium uergentem.

[p. 40]

DE BVBONE.

Βύας[2], *bubo, Anglicè, alyke foule, Germanicè* eyn ſchuffauſz / eyn ſchüffel / eyn kautz.

Aristoteles[3].

Bubo è noctuarum genere eſt, & noctuæ ſpecie quidem ſimilis, ſed magnitudine non minor quàm aquila.

Plinius[4].

Bubo funebris, & maximè abominatus, publicis præcipuè auſpicijs, deſerta incolit, nec tantùm deſolata, ſed etiam inacceſſa: noctis monſtrum, nec cantu aliquo uocalis, ſed gemitu. Volat nūquam quò libuit, ſed tranſuerſus aufertur. Hęc Plinius.

Hanc auem ſemel Venetijs iuſta aquilæ magnitudine uidi, ſed crura erant paulò breuiora quàm aquilæ crura ſolent eſſe. Cætera aquilæ ſimilis erat.

[1] Aristotle has ἑκατέρῳ ἴδιόν τι ὑπῆρχεν οὐδὲν according to one text. He has no word to represent 'appellationis.'

[2] Or Βρύας. [3] *Hist. An.* Bk VIII. 39.

[4] *Hist. Nat.* Lib. X. cap. xii.

there any difference between the two, save that of colour and of voice. That the bird is the same is evident, since, while this kind is undergoing the change, each of the two is seen, not fully changed as yet, and neither having so far any proper name. Thus far Aristotle.

In England I have never seen the Atricapilla, nor yet but once in life, and that in Italy within the house of the Duke of Ferrara. And Don Francisco of the Holy Council of the Duke, a man uncommonly well versed in both the tongues, shewed it to me himself. The bird was like the English lingett and the German grasmuschen, so far as size of body went; but it had a black head, and the remaining colour of its body verging more to grey.

Of the Bubo.

Βύας, bubo, in English alyke foule, in German eyn schuffauss (eyn schüffel), eyn kautz.

Aristotle.

The Bubo is of the race of Noctuæ, and is in aspect very like a Noctua, but not less than an Aquila in size.

Pliny.

The Bubo is a fatal bird, of evil omen beyond other sorts, especially at public auguries; it lives in desert places, and not merely those that are unpeopled, but those also hard of access: monster of the night it utters not a song, but only a groan. It never flies where it intends, but is borne off aslant. So Pliny says.

This bird I saw at Venice once, of a full Eagle's size[1], its legs, however, were a little shorter than an Eagle's legs are wont to be. But it was like an Eagle otherwise.

[1] Turner probably meant the Eagle Owl (*Bubo ignavus*).

[p. 41] *DE BOSCA.*

Boſca, auis eſt aquatica, anati ſimilis, ſed minor. Quum multæ ſint aues aquaticæ anati ſimiles, ſed minores, ut ſunt, telæ uocatæ ab Anglis Vuigene & pochardæ, eam puto boſcam eſſe, quæ proximè ad magnitudinem & ſimilitudinem anatis accedit. Hoc quum pocharda faciat, illam Ariſtotelis eſſe boſcam iudico.

DE CAPRIMVLGO.

Caprimulgus, αἰγοθήλας.

Aristoteles[1].

Caprimulgus, auis eſt montana, magnitudine paulò maior, quàm merula, minor quàm cuculus, moribus mollior. Parit duo oua, aut tria cùm plurima. Sugit, caprarum ubera aduolans, unde nomen accepit. Cùm uber ſuxerit, extingui, capramq; excæcari aiunt, parùm clarè interdiu uidet, ſed noctu perſpicax.

Plinius[2].

[p. 42] Caprimulgi appellantur grandiores merulæ aſpectu, fures nocturni, interdiu etiam uiſu carent.

Cùm eſſem apud Heluetios, ſenem quendam conſpicatus, capras paſcentem in montibus, quos herbas quærendi gratia aſcenderam, rogabam num auem nouiſſet merulæ magnitudine, interdiu cæcam, noctu perſpicacem, quæ caprarum ubera noctu ſugere ſoleat, unde capræ poſtea cæcæ euadunt? qui reſpōdit, ſe in Heluetiorum montibus ante quatuordecim annos, multos uidiſſe, & multas iniurias ab ipſis paſſum, ut qui ſemel

[1] *Hist. An.* Bk IX. 109.

[2] *Hist. Nat.* Lib. X. cap. xl.

OF THE BOSCA.

The Bosca is a water fowl, like to a Duck[1], but smaller. Since there are many water fowls like to the Duck, but smaller (as there surely are), called Teles, Wigenes and Pochards by the English, I believe the Bosca to be that which comes nearest the size and likeness of a Duck. And since the Pochard does so, I decide that it is Aristotle's Bosca.

OF THE CAPRIMULGUS.

Caprimulgus, *αἰγοθήλας*.

ARISTOTLE.

The Caprimulgus is a mountain bird, in size a little bigger than the Merula, less than the Cuculus; in disposition it is milder. It lays two eggs or three at most. Flying to the udders of she-goats, it sucks them, and thus gets its name. They say that the udder withers[2] when it has sucked at it, and that the goat goes blind. By day the bird sees dimly, but quite well at night.

PLINY.

Caprimulgi, as they are called, look bigger than a Merula, and act as thieves by night; by day they even lack the power of sight.

When I was in Switzerland I saw an aged man, who fed his goats upon the mountains, which I had gone up intent on search of plants: I asked him whether he knew a bird of the size of a Merula, blind in the day-time, keen of sight at night, which in the dark is wont to suck goats' udders, so that afterwards the animals go blind. Now he replied that he himself had seen many in the Swiss mountains fourteen years before, that he had suffered many losses from those very

[1] That is, the Wild Duck (*Anas boscas*).

[2] This rendering appears much preferable to translating this word 'the goat dies,' as, judging from the punctuation, some would have it.

ſex capellas à caprimulgo occæcatas habuerat. cæterū nūc omnes ad unum ab Heluetijs uſque ad inferiores Germanos, ubi hodie non ſolùm capras lacte priuant & occæcant, ſed & oues inſuper occidunt, auolaſſe. Nomen auis quærenti, paphum, id eſt, ſacerdotem dici reſpōdit. Sed uetulus ille mecū fortè iocatus eſt. Ego uerò, ſiue iocatus fuerit, ſiue ſeriò locutus, aliud Germanicū caprimulgi nomen quā quod me docuit iſte, non teneo. Si qui ſint, qui melius & aptius nomen in prōptu habeant, proferant.

DE CARDVELE.

Carduelis, ſi Gazæ credimus, Græcè θραυπίς *dicitur,* [p. 43] *& inter ſpiniuoras auiculas Ariſtoteles recenſet. Nec plura de carduele apud Ariſtotelem lego. Plinius*[1] *ſcribit cardueles auium minimas imperata facere, nec uoce tantùm, ſed pedibus & ore pro manibus.*

Præter auiculam illam ſpiniuoram aurea uitta redimitam, aliam noui ſpiniuoram colore uiridem, quæ non ſecus atq; auriuittis roſtro è duabus ſitulis uiciſſim aſcendentibus & deſcendentibus, cibum ex una, & potū ex altera deſumit. Quin & hoc facit miliaria, quam linotam noſtrates appellant. Eadē quoq; hominē quiduis cātantē, uoce imitatur. Quare nō ſola illa, quæ Græcè θραυπίς, *& Latinè Theodoro carduelis dicitur, imperata facit, & roſtro & pedibus pro manibus utitur.*

Dictu mihi difficile uidetur, quam'nam è tribus, quum omnes illæ carduorum ſemine ueſcantur, Plinius carduelem fecerit, nū thraupin aut acanthin, aut chry-

[1] *Hist. Nat.* Lib. x. cap. xlii.

birds; so that he had once had six she-goats blinded by Caprimulgi, but that one and all they now had flown away from Switzerland to Lower Germany, where nowadays they did not only steal the milk of she-goats, making them go blind, but killed the sheep besides. And, on my asking the bird's name, he said that it was called the Paphus, otherwise the Priest. But possibly that aged man was jesting with me. Yet whether he was jesting, or spoke gravely, still I have no other German name than what he gave to me for Caprimulgus. If there be any then who have in readiness a better or a fitter name than this, let them produce it.

Of the Carduelis.

The Carduelis, if we believe Gaza, is in Greek called θραυπίς. Aristotle also numbers it among small thistle-eaters[1]. I find in Aristotle nothing more than this about the Carduelis. Pliny writes that Cardueles, smallest of all birds, perform set tasks, and not in song alone, but with their feet and beak in place of hands.

Besides that thistle-eating little bird[2] adorned with band of gold I know another thistle-eating sort, in colour green[3], which with its beak takes up its food from one of two small buckets moving up and down alternately, its water from the other, as the Aurivittis does. The Miliaria moreover does the same, which bird our countrymen call Linot. Furthermore it mimics with its song a man when singing anything. And so it is not only that one kind, in Greek called θραυπίς and in Latin named by Theodorus[4] Carduelis that performs the tasks that it is bid, and uses beak and feet in place of hands.

It seems to me then difficult to say, since all three birds feed upon thistle-seeds, which of them Pliny meant by Carduelis, whether it should be the Thraupis, or the Acanthis, or the Chrysomitris. And should it be the Thraupis, as

[1] See p. 35.

[2] The Aurivittis, p. 35.

[3] Probably Turner means the Siskin (*Carduelis spinus*).

[4] Theodorus Gaza.

ſomitrẽ. Si thraupin, ut Gaza credit, auis illa aureis plumis carduelis non erit: nam illa apud Ariſtot. nõ θραυπίς, *ſed Chryſomitris eſt. Quæ'nã igitur auis carduelis ſit, non audeo pronunciare.*

DE CŒRVLEONE.

Κυανός, *cœruleo, Anglicè, a clotburd, a ſmatche, an arlyng, a ſteinchek, German.* eyn brech uögel.

[p. 44]

Aristoteles[1].

Cœruleo maximè in Scyro[2] colit, ſaxa amans: magnitudine minor quàm merula, maior paulò quã fringilla: pede magno[3] eſt, ſcanditq; ſaxa: colore cœruleo: roſtro tenui & longo: crure breui, ſimiliter ut pipo eſt.

Cœruleo, ſi ea ſit auis, quam conijcio eſſe, in cuniculorum foueis, & ſub lapidibus in Anglia nidulatur, & in hyeme non apparet.

DE CERTHIA.

Aristoteles[4].

Certhia, auicula eſt exigua, cui mores audaces, domicilium apud arbores, uictus ex coſſis, ingenium ſagax in uitæ officijs.

Auis eſt quædam, quam Angli creperam, id eſt, reptitatricem nominant, quòd ſuper arbores ſemper repat, quam certhiam eſſe credo. Ea regulo paulò maior, pectore pallido, cætera fuſca & maculis nigris diſtincta

[1] *Hist. An.* Bk IX. 97.
[2] Another reading is Νισύρῳ.
[3] μεγαλόπους, but some read μελανόπους.
[4] *Hist. An.* Bk IX. 92.

Gaza believes, that bird with golden feathers will not be the Carduelis, for with Aristotle that is not the *θραυπίς*, but the Chrysomitris. Therefore I dare not pronounce what bird the Carduelis is.

OF THE CŒRULEO.

Κυανός, cœruleo, in English a clot-burd, a smatche, an arlyng, a steinchek, in German eyn brech vögel.

ARISTOTLE.

The Cœruleo chiefly dwells in Scyros and loves rocks; in size it is a little smaller than a Merula, a little larger than a Fringilla: it has large feet, and climbs on rocks: in colour it is blue: the beak is long and thin: the legs are short as in the Pipo.

The Cœruleo[1], if it be the bird which I conjecture, nests in rabbit holes and under stones in England, and does not appear in winter.

OF THE CERTHIA.

ARISTOTLE.

The Certhia is a very little bird of bold habits; its home is upon trees, its food is grubs; it shews wise instinct for the needs of life.

There is a certain bird which Englishmen call Creeper, that is Climber, for it always climbs about on trees: this I believe to be the Certhia. It is a little bigger than the Regulus, having a whitish breast, the other parts dull brown, but varied with black spots; its note is sharp, its

[1] Whatever bird Aristotle's may be, Turner's is certainly the Wheatear. Belon and Gesner seem to think that the former is the Blue Thrush, but Sundevall is certainly wrong in suggesting that it is the Wall-Creeper (*Tichodroma muraria*).

[p. 45] *eſt, uoce acuta eſt, & roſtro tenui, & leuiter in fine adunco, nunquam quieſcit, ſed ſemper per arbores picorum more ſcandit, & coſſos è corticibus eruens, comedit.*

DE CICONIA.

Πελαργός, *ciconia, Anglicè a ſtork, Germanicè* eyn ſtorck, *& Saxonicè* eyn ebeher.

Ciconia, ut Germanis auis eſt notiſſima, ita Britannis meis pleriſq; omnibus tam ignota eſt, quàm quæ omnium ignotiſſima. Nec mirum, quum nuſquam in inſula noſtra niſi captiua Ciconia uideatur. Apud Germanos in ſummis tectis, aliquando in ipſis ſummis fumarijs nidulatur. Auis eſt mediæ magnitudinis inter gruem & ardeam, pennis albis & nigris diſtincta: crura longa habet, roſtrum gruinam, ſed rubrum & craſſum: circa lacus & paludes degit, ranas, bufones, angues, & piſces comedens.

Plinius de ciconiis[1].

Ciconiæ, quo'nam è loco ueniant, aut quò ſe referant, incompertum adhuc eſt. E longinquo uenire, non dubium, eodem, quo grues modo, illas hyemis, iſtas æſtatis aduenas. Abi- [p. 46] turæ congregantur in loco certo: comitatæq; ſic ut nulla ſui generis relinquatur, niſi captiua & ſerua, ceu lege prædicta die recedũt. Nemo uidit agmen diſcedentium, cùm diſceſſurum appareat, nec uenire, ſed ueniſſe cernimus. Vtrumque nocturnis fit temporibus. Ciconiæ nidos eoſdem repetunt, & genetricum ſenectutem educant.

DE CINCLO.

Κίγκλος, ἡ σεισοπυγίς, *Anglicè a uuater ſuuallouu, Germanicè* eyn ſteynbiſſer.

[1] *Hist. Nat.* Lib. x. cap. xxiii.

beak is slender and is slightly hooked towards the tip; it never rests, but is for ever climbing up the trunks of trees after the manner of the Woodpeckers, and it eats grubs, picking them from the bark.

Of the Ciconia.

Πελαργός, ciconia, in English a stork, in German eyn storck, and in Saxon eyn ebeher.

The Stork, though one of the best known of birds among the Germans, is to nearly all my countrymen of Britain as unknown as the most unknown bird. And little wonder since a Stork is nowhere to be seen, save as a captive, in our island. With the Germans the bird nests upon roofs or even chimney tops at times. It is a bird of middle size between a Crane and a Heron and distinguished by feathers of black and white: it has long legs and a beak like a Crane's, but red and stout; it lives near lakes and marshes, eating frogs, toads, snakes and fishes.

Pliny on Ciconiæ.

From what parts the Ciconiæ may come, or whither they betake themselves, is not yet ascertained. It is indubitable that they come from far in the same manner as the Grues, but the former arrive in winter and the latter in summer. When ready to depart these birds collect at some fixed place, and after gathering, so that none of their tribe, unless a prisoner or a slave, is left behind, they disappear, on an appointed day, as if by law. No one has ever seen the whole array in very act to go, though it may haply shew itself when ready to depart; nor do we view it coming, but when it has come. Both these events take place at night. Ciconiæ seek the same nests again, and tend their parents in old age.

Of the Cinclus.

Κίγκλος, ἡ σεισοπυγίς, in English a water swallow, in German eyn steynbisser.

ARISTOTELES[1].

Cinclus ex mari & fluuijs uictũ petit. aſtutus eſt, & captu difficilis: ſed captus, omnium maximè miteſcit. Lęſus hic eſt, incontinens enim parte ſui poſteriore.

Auicula, quã ego cinclũ eſſe puto, galerita paulò maior eſt, colore in tergo nigro, uentre albo, tibijs longis, & roſtro neutiquam breui: uere circa ripas [p. 47] *fluminum, ualde clamoſa eſt & querula, breues & crebros facit uolatus.*

DE CHALCIDE.

ARISTOTELES[2].

Chalcis rarò apparet, montes etenim incolit. nigro colore eſt, magnitudine accipitris, quem palumbarium nominant: forma longa ac tenui, Iones cymindem appellant. cuius Homerus etiam meminit in Iliade cùm dicit:

Chalcida dij perhibens, homines dixêre cymindem[3].

Sunt, qui eandem hanc auẽ non aliam eſſe atque ptyngem uelint. Interdiu minùs apparet, quia non clarè uidet, ſed noctu uenatur, more aquilæ: pugnat uerò cum aquila adeò acriter, ut ſæpius ambæ implexæ, deferantur in terram[4], & uiuę à paſtoribus capiantur. Parit hęc oua duo, & ſaxis ſpeluncisq; nidulatur.

[1] *Hist. An.* Bk IX. 76.

[2] *Hist. An.* Bk IX. 79—80.

[3] *Iliad*, Bk XIV. l. 291. 'Perhibens' is a misprint for 'perhibent.'

[4] Aristotle has not these five words.

ARISTOTLE.

The Cinclus seeks its food from the sea and from rivers. It is cunning and is hard to catch, but grows the gentlest of all birds when caught. It is moreover maimed, being without control over its hinder parts.

The little bird which I believe to be the Cinclus[1] is a little bigger than the Galerita, with black colour on the back, and a white belly; while it has long shanks and a bill by no means short: in spring it is exceeding clamorous and querulous about the banks of rivers, where it takes short and incessant flights.

OF THE CHALCIS.

ARISTOTLE.

The Chalcis is not often seen, for it haunts mountains. It is of a black colour, and of the size of the Accipiter which they name Palumbarius. Its form is long and slender; the Ionians call it Cymindis. Of it furthermore Homer makes mention in the Iliad, wherein he says:—

The Gods know it as Chalcis, men say Cymindis.

Some there are who would make this very bird none other than the Ptynx. By day it shews itself but little, since it does not see clearly, although it hunts by night, after the manner of the Aquila; yet with the Aquila so keenly does it strive, that frequently both birds fall to the ground clutching each other, and are caught alive by shepherds. It lays two eggs, and nests in cliffs and caves[2].

[1] Turner evidently means the Common Sandpiper (*Actitis hypoleucus*).

[2] Sundevall says that Külb thought that this bird was the Hawk Owl, but himself refers it doubtfully to the Glossy Ibis (*Plegadis falcinellus*), which however is not a night bird and does not live on hills. Belon and Gaza thought that it was an owl of some kind.

[p. 48]

DE COLLVRIONE.

Κολλυρίων, *collurio, Anglicè, a feldfare aut a feldefare. Quibusdã German.* eyn krammesuögel.

ARISTOTELES [1].

Collurio ijsdem, quibus merula uescitur, magnitudo eius eadem quæ superioribus, id est, uireoni & mollicipiti [2], capitur potius hyberno tempore.

Auis, quam collurionem esse puto, turdum magnitudine æquat, sed caudam habet longiorem, & magis mobilem, & pectus maculosum. In æstate apud nos aut rarò aut nunquam uidetur: in hyeme uerò tanta copia est, ut nullius auis maior sit. baccis aquifoliæ arboris, sorbi minimæ, & similium arborum uescitur. gregatim uolat, & inter uolandum obstrepera est.

DE COLVMBIS.

Περιστερά, *columba, Anglicè a doue, Germanicè* eyn taube. *Saxonicè* eyn duue.

ARISTOTELES [3].

Columbacei uerò generis plures species sunt. Est enim liuia à liuore dicta, diuersum certè à columba genus, quippe minor quàm columba sit, & minùs patiens mansuescere: liuet enim plumis, & penè nigricat, & pedibus rubris scabrosisq; est. Quas ob res, nullus id genus cellare alit. Maximo inter hęc genera sunt corpore palumbes. Secundum magnitudinis locum uinago obtinet, paulò maior quàm columbus [4] est. Minimum ex his turtur est. pariunt columbæ omnibus anni temporibus, pullosq; educant si locum apricum habeãt & cibum. Sin minùs,

Liuia. [p. 49] Palumbes. Vinago. Turtur.

[1] *Hist. An.* Bk IX. 99.

[2] These five words are not represented in Aristotle. Moreover the πάρδαλος is here omitted, and is placed later (p. 107 of the original work), but there seems to be some doubt as to the correct reading in the Greek.

[3] *Hist. An.* Bk V. 43. The readings vary in places, but the rendering is decidedly free.

[4] Gaza and Turner make indiscriminate use of 'Columbus' and 'Columba' for the same kind of bird.

OF THE COLLURIO.

Κολλυρίων, collurio, in English a feldfare or a feldefare. According to some Germans eyn krammesvogel.

ARISTOTLE.

The Collurio feeds on the same meat as the Merula. Its size is that of the aforesaid kinds—that is, the Vireo and the Molliceps—it is caught chiefly in the winter time.

The bird which I consider to be the Collurio equals a Thrush in size, but has a longer and more flirting tail, also a spotted breast. Rarely or never is it seen with us in summer: yet its plenty is so great in winter that of no kind is there more. It eats the berries of the Holly, the Least Service, and like trees. It flies in companies, and on its flight is very noisy.

OF DOVES.

Περιστερὰ, columba, in English a dove, in German eyn taube, in Saxon eyn duve.

ARISTOTLE.

Of the Dove-kind, however, there are many sorts. For first there is the Livia, named from its livid colour, which is certainly a different kind from the Columba, inasmuch as it is smaller and less ready to be tamed: and it is livid in its plumage, verging upon black, and has moreover red and roughened feet. Wherefore nobody keeps this kind in cotes. Of greatest size among the several sorts are the Palumbes; the Vinago holds the second place herein, a little bigger than the Columbus. The smallest of them is the Turtur. The Columbæ breed at all times of the year, and rear their young, if they have but a sunny place and food. If otherwise they breed

æſtate tantummodo fœtant. Sed proles præſtantior uere eſt, quàm autumno, deterrima æſtate, & omni tempore calidiore.

De ijſdem in alio libro[1] *ad hunc modum ſcribit,*

Alia frugibus uiuunt, ut palumbes, co- [p. 50] lumbus, uinago, turtur. Viſuntur ſemper columbi, atque palumbes, ſed turtur æſtate tantũ, hyeme ſe condit, latitat[2] enim ſuo tempore. Vinago autumno & conſpicitur, & capitur, cui magnitudo maior, quàm columbo, minor quã palumbi eſt.

Liuia. Πελειάς, *quæ liuia Latinè dicitur, eſt ſylueſtris illa columba, quam Angli a ſtocdoue, & Germani* eyn holtz taube *nominant.*

Palumbes. Φάττα, *Latinè palumbes ſiue palumbus, dicta ab Anglis a coushot a ringged doue, & à Germanis* eyn ringel taube *appellatur. Hæc longè aliter atque liuia nidulatur. Nidificat autem liuia in cauis aliquando arboribus, interdum & in templorum muris. Palumbes uerò in condenſa hedera, aut ſuper ramum arboris ex pauculis ligniculis tranſuerſim poſitis, tenuiſſimum nidum conſtruit. Quòd ſi quis mihi parùm hac in re fidat, torquatos columbos esse palumbes ueterum, Ariſtotelem diligentiùs legat, & Martialem*[3] *poëtam de ijſdem ita ſcribentem audiat,*

[p. 51] *Inguina torquati, tardant hebetantq; palumbi*
Non edat hanc uolucrem, qui uolet eſſe ſalax.

Politianus[4] *de palumbis ita ſcribit:*

Dum ſua torquatæ repetunt dictata palumbes.

Τρυγών. Turtur. *Turtures in Germania ſunt multò frequentiores quã in Anglia. Turturem Angli & Saxones communi uocabulo* turtel duue *nominant.*

[1] *Hist. An.* Bk VIII. p. 45.
[2] The Greek is *φωλεῖ γάρ*. But how far Aristotle referred to birds 'hibernating' is very doubtful.
[3] *Epigr.* Lib. XIII. lxvii.
[4] A poet and scholar of the Renaissance.

only in summer. Yet in spring the young are better than in autumn, they are worst of all in summer and at every hotter season.

Of the same he writes in another book after this manner:—

Other birds live on crops, as the Palumbes, the Columbus, the Vinago and the Turtur. The Columbi and Palumbes may be always seen, the Turtur only in the summer. In the winter it lies hid, for it conceals itself at the due time. But the Vinago is both seen and caught in autumn, of which bird the size is greater than that of the Columbus, but less than that of the Palumbes.

Πελειάς, which in Latin is called Livia, is that dove of the woodlands which the English name a stocdove, and the Germans eyn holtztaube.

Φάττα, in Latin Palumbes or Palumbus, is called by Englishmen a Coushot or a Ringged Dove, and by Germans named eyn ringel taube. It nests far otherwise than does the Livia, for that bird sometimes breeds in hollow trees and sometimes even in the walls of churches. But the Palumbes builds a nest of the frailest of a few small twigs laid crosswise in a mass of ivy or upon a bough. Now in this thing if there be anyone who places little confidence in my opinion that our collared doves are the Palumbes of the ancients, let him read with greater care his Aristotle and give ear as well to Martial the poet writing thus of the same birds:—

> Ringed doves make a man's loins slow and dull;
> Who would be lusty should not eat this bird.

Politian writes of the Palumbi thus:—

> While ringed doves seek again their accustomed haunts.

Turtle Doves are much more plentiful in Germany than in England. English and Saxons in common call it turtel duve.

Vinago. Οἰνάς, *quæ Latinè uinago dicitur, mihi nunquã hactenus uiſa eſt, nec quid habeat nominis apud noſtros, aut apud Germanos compertum habeo. Vidi tamen Venetijs columbos hijs noſtratibus ſeſquialtera portione maiores: ſed hos non uinagines fuiſſe credo, ſed columbos è Campania ad Venetos aduectos, ubi Plinius columbos ſcribit eſſe grandiſſimos.*

DE COTVRNICE.

ὄρτυξ, *coturnix, Anglicè a quale, Germanicè* eyn wachtel.

Plinius[1].

Coturnix parua auis, & cùm ad nos uenit, terreſtris potiùs quàm ſublimis. Aduolant & hę ſimili modo, quo grues & ciconiæ, non ſine [p. 52] periculo nauigantium, cùm appropinquauere terris. Quippe uelis ſæpè inſidunt, & hoc ſemper noctu, merguntq; nauigia. Coturnicibus, ueratri, ſiue ut alij legunt, ueneni: ſemen gratiſſimus cibus. quam ob cauſam eas damnauere menſæ, ſimulq; comitialẽ propter morbum deſpui ſuetum, quem solę animaliũ ſentiunt præter hominem.

Quæ, quum ita ſe habeant, demiror quis malus genius Britannis meis in mentem immiſit, ut eas tantopere in deliciis habeant, quum tot malis, ueneno ſcilicet, et comitiali morbo, illarũ caro ſit obnoxia. Coturnix perdici ſimilis eſt: ſed multis partibus minor. Coturnix, ut ſcribit Ariſtot. hoc ſibi peculiare & proprium uindicat, ut & ingluuiem, & gulam propè uentriculum amplam & latam habeat.

[1] *Hist. Nat.* Lib. x. cap. xxiii.

Οἰνάς, in Latin called Vinago, has never met my eye up to this time, nor have I yet found out what name it bears among our countrymen or among Germans. But I have seen doves in Venice half as big again as those of our own land, although I do not think that they could be Vinagines, but birds brought to those parts out of Campania, where Pliny notes the Doves to be exceeding large.

OF THE COTURNIX.

ὄρτυξ, coturnix, in English a quale, in German eyn wachtel.

PLINY.

The Coturnix is a little bird, and, when it comes to us, keeps on the ground more than aloft. Yet it flies hither just as Grues and Ciconiæ, not without danger to sea-faring men, when they approach the land. For these birds often settle on the sails, and that always at night, and so sink ships. The seed of Veratrum, or, as others read, Venenum, is a very grateful food to the Coturnices, and for this cause men have condemned them for the table; furthermore it is the custom for them to be spurned on account of the falling sickness, to which, they alone of animals, save man, are subject.

Now since these things are so, I marvel much what evil genius put it into the mind of my fellow Britons to esteem them thus among their delicacies, when their flesh is liable to ills so many, namely poison and the falling sickness. The Quail is like the Partridge, although many times smaller. As Aristotle writes, it claims a property peculiar to it of having both crop and gullet large and wide near to the stomach.

DE CORNICE.

Κορώνη, *cornix, Anglicè a crouu, Germanicè* eyn krae, *&* eyn kraeg. *Cornix auis eſt omniuora, nam carnes, piſces, & grana interdum uorat, circa littora maris, & ripas fluminum multùm uerſatur, ut ea animalia, quæ* [p. 53] *unda eiecit, tangat. Cornix tota nigra eſt. & media magnitudine inter monedulam & coruum.*

Eſt & marina quædam cornix, quam aliqui hybernam cornicem uocant, capite, cauda, & alis nigris, cætera cineria: an hanc aliquando uiderint Ariſtoteles & Plinius, dubito: nam de ea nuſquam mentionem fecerunt. Supereſt adhuc & alia cornix graniuora, roſtro albo, cætera nigra. Hãc σπερμολόγον, *id eſt, frugilegam Ariſtotelis Longolius eſſe coniecit.*

DE CORVO.

Κόραξ, *coruus, Anglicè a rauen, Germanicè* eyn rabe. *Coruus, quum ſit auis cornice maior, tota nigra & carniuora, omnibus ſatis notus eſt. Corui locis arctioribus & ubi ſatis pluribus non ſit, duo tãtùm incolunt, & ſuos pullos cùm iam poteſtas uolandi eſt, primùm nido eijciunt, deinde regione tota expellunt. Parit coruus quatuor aut quinq;.*

DE CVLICILEGA.

Κνιπολόγος, *culicilega, Anglicè a uuagtale, Germanicè* eyn waſſer ſteltz. eyn quikſtertz.

ARISTOTELES [1].

Culicilega, magnitudine eſt quãta ſpinus, [p. 54] colore cinerea, diſtincta maculis, uoce parua, quæ & ipſa lignapeta [2] eſt.

Culicilegam eſſe iudico auiculam, quam aliqui motacillam nuncupant. eſt autem illa albo & nigro uariè

[1] *Hist. An.* Bk VIII. 44.

[2] The Greek is ἔστι δὲ καὶ τοῦτο ξυλοκόπον.

OF THE CORNIX.

Κορώνη, cornix, in English a crow, in German eyn krae and eyn kraeg.

The Crow[1] is an omnivorous bird, for it eats flesh and fish and sometimes grain; it much frequents sea-coasts and river-banks, that it may there obtain those animals which the tide has thrown up. The Crow is wholly black and is midway in size between a Daw and a Raven.

There also is a certain Sea Crow, which some call the Winter Crow[2], with black head, tail, and wings and the remainder grey: but whether Aristotle or Pliny ever saw this bird I am uncertain, for they have not mentioned it in any place. There still remains another Crow[3], a grain-eater, with white beak, but black otherwise. Longolius conjectured this to be Aristotle's σπερμολόγος, that is frugilega.

OF THE CORVUS.

Κόραξ, corvus, in English a raven, in German eyn rabe.

The Raven, inasmuch as it is bigger than the Crow, quite black, and a flesh-eater, is sufficiently well known to all. In places with less space, and where there is not room for many, Ravens dwell only in pairs, and, when their young have just gained power of flight, the parents first banish them from the nest, and later drive them out of the whole neighbourhood. The Raven has a brood of four or five.

OF THE CULICILEGA.

Κνιπολόγος[4], culicilega, in English a wagtale, in German eyn wasser steltz, eyn quikstertz.

ARISTOTLE.

The Culicilega is a bird of the same size as the Spinus, ash-coloured, and marked with spots: its voice is poor; moreover it pecks wood.

The Culicilega I judge to be that little bird, which some name Motacilla, inasmuch as it is variously marked with

[1] The Carrion Crow (*Corvus corone*).
[2] The Hooded or Grey Crow (*Corvus cornix*).
[3] The Rook (*Corvus frugilegus*).
[4] Sundevall thinks that this bird is *Certhia familiaris*.

diſtincta, cauda longa, quam ſemper motitat. degit plurimum ad ripas fluminum, ubi muſcas captat & uermiculos, quin & aratrum uermium cauſa ſequitur, quos uerſat & exhibet cum gleba aratrum.

DE CVCVLO.

Κόκκυξ, *cuculus, Anglicè a cukkouu, & a gouke, Germanicè* eyn kukkuck.

Aristoteles[1].

Cuculus ex accipitre fieri, immutata figura, à nonnullis putatur: quoniam quo tempore is apparet, accipiter ille, cui ſimilis eſt, non aſpicitur. Sed ita ferè euenit, ut ne cæteri item accipitres cernãtur cùm primam uocem emiſit cuculus, niſi perquàm paucis diebus. Ipſe autem breui tempore ęſtatis uiſus, hyeme nõ [p. 55] cernitur. Eſt hic neque aduncis unguibus, ut accipiter, neq; capite accipitri ſimilis: ſed ea utraque parte columbum potiùs quàm accipitrem repræſentat. Nec alio quàm colore imitatur accipitrem, niſi quòd maculis diſtinguitur, ceu lineis, cuculus uelut punctis. Magnitudo atq; uolatus ſimilis accipitrũ minimo, qui magna ex parte non cernitur per id tempus, quo cuculus apparet. Nã uel ambo unà uiſi aliquando ſunt. Quin etiam ab accipitre interimi[2] cuculus uiſus eſt, quod nulla auis ſuo in genere ſolet facere. pullos cuculi nemo ait ſe uidiſſe. parit tamẽ, uerùm non in nido, quem ipſe fecerit, ſed

[1] *Hist. An.* Bk VI. 41—44.
[2] Aristotle has κατεσθιόμενος.

black and white, and it has a long tail, which it is always jerking. It mostly haunts the banks of rivers, where it catches flies and little worms; moreover it follows the plough for the sake of the worms which are turned up and laid bare with the clod.

Of the Cuculus.

Κόκκυξ, cuculus, in English a cukkow, and a gouke, in German eyn kukkuck.

Aristotle.

By some the Cuculus is thought to come by change of form from an Accipiter[1], since, at the season when the former appears, the Accipiter which it resembles is not seen. But commonly it so falls out that the other Accipitres are likewise absent when the Cuculus utters its earliest cry, save for a very few days. Further the bird itself is only seen for a short time in summer; it is not observed in winter. Nor has it the claws hooked as an Accipiter, nor yet a head like an Accipiter: but in both of these parts it counterfeits a Columbus rather than an Accipiter. In naught but colour does it imitate the Accipiter, except that in its marks, it is distinguished as it were by lines, the Cuculus by spots. The size and mode of flight are like those of the least of the Accipitres, which for the most part at the time wherein the Cuculus appears, is not to be observed. Yet on occasion both have been seen at once. The Cuculus, moreover, has been known to be struck down by the Accipiter, which thing no bird is ever wont to do to one of its own kind. Nobody says that he has seen young of the Cuculus, and yet it breeds, although not in a nest which it has made itself: but sometimes

[1] Such a tradition is still common in many parts of this country and on the Continent.

interdum in nidis minorum auium, & oua, quæ aliena reperit, edit : maximè uerò nidos palũ- [p. 56] bium petit, quorum & ipſorum oua eſu abſumit, ſua relinquens : parit maiori ex parte ſingula oua, rarò bina. Curucæ quoque in nido parit, fouet illa & excudit & educat. Quo quidem præcipuè tempore[1] & pinguis & grati ſaporis pullus cuculi eſt. Genus eorum quoddam nidos facere procul in petris excelſis, præruptisq; aſſolet.

Curuca a titlyng.

Cuculum hîc nobis ſatis graphicè depinxit Ariſtoteles, ſi curucam eadem diligentia deſcripſiſſet, non fuiſſet hodie tam omnibus ferè incognita quã nunc eſt. Ego ſuſpicor Anglorum titlingam eſſe curucam Ariſtotelis. Nam nullam auem in uita frequentiùs cuculi pullum ſequentem, & pro ſuo educantem, quam illam obſeruaui. Eſt autem illa luſcinia minor, ſed eadem corporis figura, colore ſubuiridi, culices & uermiculos in ramis arborum ſectatur, rarò humi conſiſtit, hyeme non cernitur.

De crece ex Aristotele[2].

[p. 57] Sed cũ omnibus quaterni digiti ſint, tres parte priore habentur, unus parte poſteriore pro calce, ut tute ſit, qui minutus ineſt ijs, quæ longa habent crura, ut in crece euenit[3]. Eſt[4] autem crex moribus pugnacibus, ingenio ualens ad uictum, ſed cætera infœlix.

Eſt auis quædam apud Anglos, longis cruribus, cætera coturnici, niſi quòd maior eſt, ſimilis, quæ in ſegete & lino, uere et in principio æſtatis non aliam

[1] This apparently means 'when in the nest.' How then does Aristotle say that 'no one has ever seen the young'? The passage may be an interpolation, as may be another which follows referring to Hawks. If so, the fact of nesting on rocks may also refer to Hawks, and be a further interpolation.

[2] *Hist. An.* Bk II. 46.

[3] 'ut...evenit.' These words are not in Aristotle.

[4] *Hist. An.* Bk IX. 91.

in the nests of smaller birds, and it devours the eggs of the others that it finds. It mostly seeks the nests of the Palumbes and eats those birds' eggs, leaving its own behind. For the most part it lays a single egg or rarely two. It also lays in the Curuca's nest, and that bird sits upon the eggs, hatches and rears the young. And at that time indeed the offspring of the Cuculus is both particularly fat and of a grateful flavour. A certain kind of Cuculus is wont to make its nest far off on steep and very lofty rocks.

Here Aristotle has portrayed the Cuculus to us most graphically, and, had he described the Curuca in the same careful way, it would not at this day have been so little known to almost everyone as now it is. The Curuca of Aristotle I suspect to be the Titling[1] of the English. For I have observed no other bird in life more frequently than this following the Cukkow's young and rearing it, as though its own. Now it is less than the Luscinia, but with the same figure of body, and in colour somewhat green; it hunts for gnats and little worms among the boughs of trees. It seldom settles on the ground, and is not seen in winter.

Of the Crex from Aristotle.

But seeing that all birds have four toes each, three are directed forwards and one backwards by way of a heel, for safety's sake; the last is very small in such as have long legs, as happens with the Crex. Further the Crex is of pugnacious habit, clever in procuring food, but of bad omen otherwise.

There is a certain bird in England with long legs, otherwise like a Quail, except that it is bigger, which in spring as well as early summer makes no other cry among the corn and flax

[1] It is impossible to say with certainty what Turner's 'Titlyng' was; but probably he meant the Tree-Pipit, which he confounded with the Titlark.

habet uocem, quàm crex crex: hãc enim uocem ſemper ingeminat, quam ego Ariſtotelis crecem eſſe puto. Angli auem illam uocant a daker hen, Germani eyn ſchryk. *nuſquam in Anglia niſi in ſola Northumbria uidi & audiui.*

De Diomedeis avibus ex Plinio[1].

Nec Diomedeas præteribo aues. Iuba cataractas uocat, & eis eſſe dẽtes, oculosq́ igneo colore, cætera cãdidos tradẽs. Duos ſemper [p. 58] ijs duces, alterum ducere agmen, alterum cogere. Scrobes excauare roſtro, inde crate cõſternere, & operire terra, quæ antè egeſta fuit, in his fœtificare. Fores binas omniũ ſcrobibus, orientem ſpectare, quibus exeant in paſcua, † occidentem, quibus redeant. Aluum exoneraturas ſubuolare ſemper, & contrario flatu. Vno hæ in loco totius orbis uiſuntur, in inſula, quam diximus nobilem Diomedis tumulo, atque delubro, contra Apulię oram, fulicarum ſimiles. Aduenas barbaros clangore infeſtant. Græcis tantùm adulantur, miro diſcrimine, uelut generi Diomedis hoc tribuẽtes, ædemq́ eius quotidie pleno gutture madẽtibus pennis perluunt.

† occaſum

DE FICEDVLA.

[p. 59] Συκαλίς, *ficedula Latinè dicta, non eſt Germanorum ſneppa, quæ locis gaudet humidis, & ſolis uer-*

[1] *Hist. Nat.* Lib. x. cap. xliv.

than crex crex, and moreover it repeats this sound incessantly; I think that it is Aristotle's Crex. This bird the English call a Daker Hen, and the Germans eyn schryk[1]. I have not seen or heard it anywhere in England, save in Northumberland alone.

Of the Aves Diomedeæ from Pliny.

And I will not omit the birds of Diomede[2], which Juba calls Cataractæ, telling us that they have teeth and fire-coloured eyes, but otherwise are white. They always have two captains, one to lead the band, the other to bring up the rear. These birds dig furrows with the beak, then cover them with wattlework, and hide this with the earth thrown out at first; in these places they breed. Each furrow has two openings, one facing east, by which they may go out towards their feeding grounds, the other facing west, by which they may return. They always flutter out to disburden the belly, and against the wind. In one place only of the whole world are they to be seen, namely that island which we have set down as famous for the tomb and shrine of Diomede, over against the shore of Apulia. They are like Fulicæ. Strangers who come there they attack with clamour, only on the Greeks they fawn, with wonderful discernment, paying as it were this tribute to the race of Diomede, and every day they purify his shrine with brimming throats and water-laden wings.

Of the Ficedula.

Συκαλίς, in Latin called ficedula[3], is not the sneppa of the Germans, which delights in wet localities, and feeds only

[1] Schlegel (*Vog. Nederl.* II. 60) says that the Dutch schriek is the Water Rail (*Rallus aquaticus*), but Turner evidently means the Corn Crake (*Crex pratensis*). Naumann (*Naturg. Vög. Deutschl.* IX. p. 496) gives Schrecke as a local name for the Corn Crake.

[2] Apparently Shearwaters of some species are meant. For the story see any work on Mythology.

[3] For the supposed change of Ficedula into Atricapilla, see p. 39.

mibus uescitur: sed auicula Germanorum grasmuscho similis, ficubus & uuis uictitãs, ut pulchrè his uersibus Martialis[1] *testatur:*

Cùm me ficus alat, & pascar dulcibus uuis,
Cur potiùs nomen non dedit uua mihi?

DE FRINGILLA.

Σπίζα, *fringilla, Anglicè a chaffinche, a sheld appel, a spink, Germanicè* eyn bůchfink.

Fringillæ, autore Aristotele, æstate tepidis locis, & hyeme, frigidis degunt, & inde puto apud Latinos nomen accepisse, quòd in frigore plures conuolantes apud nos cernantur, quàm æstate. Pascerem magnitudine æquat, uarijs coloribus, albo nempe, uiridi & ruffo distincta est. maris pectus rubescit, fœminæ pallescit: cantat mas primo uere. Nidulatur fringilla in summis fruticum ramis, aut arborum infimis, nidumq; intus ex lana, forisq; ex musco facit.

DE MONTIFRINGILLA.

Οροσπίζης, *mõtifringilla, Anglicè a bramlyng, Germanicè* eyn rowert.

ARISTOTELES[2].

[p. 60] Montifringilla fringillæ similis est, & magnitudine proxima: sed collo cœruleo est, & in montibus degit, unde nomen accepit.

Auicula, quam ego montifringillam esse credo, fringillæ magnitudine & corporis figura similis est: sed mas in collo plumas habet cœruleas, quas nõ æquè promptè in fœmina depræhendas. Rostrum luteum est, & alæ uarijs coloribus, albo, nigro, & luteo nimirum distinguũtur, ut auriuittis. Vox illi insuauis & stridula est.

[1] *Epig.* Lib. XIII. xlix.
[2] *Hist. An.* Bk VIII. 41.

on worms[1]; but is a little bird like the grasmusch of the Germans, living upon figs and grapes, as Martial prettily bears witness in these lines:—

> Since the fig gives me nourishment, and I feed on sweet grapes,
> Why has the grape not rather given me a name?

Of the Fringilla.

Σπίζα, fringilla, in English a chaffinche, a sheld-appel[2], a spink, in German eyn bůchfink.

Fringillæ—on Aristotle's authority—in summer haunt warm places, and in winter cold; and thence I think that they received their name among the Latins[3], for when it is cold more are seen flocking round us than in summer time. In size the bird equals a Sparrow and is marked with various colours, namely, white and green, and russet. In the male the breast is ruddy, in the female pale. The male sings in the early spring. The Fringilla nests upon the highest boughs of shrubs or on the lowest boughs of trees, and fashions its nest inwardly of wool and outwardly of moss.

Of the Montifringilla.

'Ὀρoσπίζης, montifringilla, in English a bramlyng, in German eyn rowert.

Aristotle.

The Montifringilla is like the Fringilla, and similar in size, but with a blue neck; and it lives in mountains, whence it has its name.

The little bird which I believe to be the Montifringilla, in size and shape of body is like the Fringilla, but the male has blue feathers upon the neck, which one cannot perceive so quickly in the hen. The beak is yellow, and the wings in truth are marked with various colours, yellow, black and white, as in the Aurivittis. Its note is unmelodious and grating.

[1] See p. 35.

[2] Shell-apple, or Apple-sheiler is still a Northumbrian name for the Chaffinch. The word 'sheld' may mean 'parti-coloured.'

[3] Here Turner's mistaken etymology (*Fringilla a frigore*) is evident.

De Floro ex Aristotele[1].

Anthos, ſiue florus, uermibus paſcitur, & magnitudo illi, quanta fringillæ eſt. uictitat circa aquas & paludes, & ei color pulcher eſt, & uita commoda, odio equum habet, pellitur enim ab equo pabulo herbæ, qua ueſcitur. Nubeculans, nec ualens oculorum acie eſt, quippe qui uocem equi imitetur, atque aduolans [p. 61] equum fuget: ſed interdum excipiatur occidaturq; ab equo. In ægithum florus tantum odium gerit, ut ne mortuarum auiũ ſanguis poſſe miſceri dicatur.

DE FVLICA.

Κέπφος, *fulica, Anglicè a uuhite ſemauu uuith. a blak cop. Germanicè,* eyn wyß mewe.

Recentiores Græci, qui poſt Ariſtotelem ſcripſerunt, larum & cepphum eandem auem fecerunt, ut Eraſmus in Adagio, λάρος κέπφος, *ex Ariſtophane*[2] *& eius interprete oſtendit. Ariſtoteles uerò duas facit diuerſas aues libro de hiſtoria animalium octauo his uerbis,* ἔστι δὲ λάρος ὁ λευκός καὶ κέπφος. *Iam qua'nam ratione autores iſtos conciliē, neſcio, niſi dicam poëtas rerum peculiares & proprias notas, & diſcrimina, philoſophis multò negligentiùs obſeruantes, aues corporis figura, natalibus, & uictus ratione ſimiles, licet manifeſtis notis differentes, eaſdem aues feciſſe, quas ſeueriores philosophi ad amuſſim omnia expendentes, in diuerſas ſpecies diſtinxerunt.*

Sed inter huius ætatis grammaticos, non minor eſt opinionum uarietas de fulica, quæ'nam illa ſit, quàm [p. 62] *inter Græcos de nomine controuerſia fuit. Sunt enim*

[1] *Hist. An.* Bk VIII. 41 and Bk IX. 18, 22, freely rendered.

[2] *Pax*, l. 1067. By 'interpreter' is evidently meant the Scholiast, who says that the Proverb is used of those who promise much, and perform little.

Of the Florus from Aristotle.

The Anthos, that is Florus, feeds on worms; its size is that of the Fringilla. It gets victual round waters and marshes; its colour is fair, and its life easy to it. It holds the Horse in hatred, inasmuch as it is driven by the Horse from the grassy pastures where it feeds. It is purblind and nowise keen of eyesight, while it imitates the neighing of the Horse, and flying at it puts the Horse to flight, yet sometimes it is caught and then killed by the Horse. The Florus has so great a hatred of the Ægithus that it is stated that the blood of these two birds, even when dead, cannot be mixed.

Of the Fulica.

Κέπφος, fulica, in English a white semaw, with a black cop, in German eyn wyss mewe.

The later Greeks, who have written after Aristotle, have made the Larus and the Cepphus the same bird, which fact Erasmus in his Proverb *λάρος κέπφος* shews, from Aristophanes and his interpreter. But Aristotle in the eighth book of his History of Animals keeps the two birds distinct, using the following words:—"There is the *λάρος* that is white, also the *κέπφος*." Now in what way to reconcile these authors I know not, unless I say that poets who observe more negligently than philosophers the peculiar properties of things, and their diversities, have made these birds the same, which are alike in form of body, breeding-time, and way of feeding, although differing in manifest respects, whereas philosophers, more strict than they, gauging all things exactly, have distinguished them as different kinds.

And yet there is not less diversity of opinion among the critics of our day about the Fulica, and what that bird may be, than there was controversy among the Greeks about its name. For there are teachers of a sort[1] in Lower

[1] 'Literatores' is here apparently used in a somewhat scornful sense.

in inferiori Germania literatores aliqui, qui fulicam kyuuittam ſuam eſſe uolunt, ex eo forſan opinionem ſuam adſtruentes, quòd apud Plinium fulicæ cirrum tribui legerint. Eſt autem Germanorum kyuuitta cornice minor, plumis ferè uiridibus, et nigris per totum dorſum et caput et collū: uentre albo, longa, & ſemper erecta in capite: criſta plumea, alis obtuſioribus, & inter uolandum magnum ſtrepitum êdentibus, unde & uannellus a barbaris dicitur: aquis uermium gratia, quibus ſolis uictitat, appropinquat, ſed ipſas non ingreditur, in planis & in locis erica conſitis, plurimùm degit. Ad depopulandum uermes, noſtrates in hortis ſæpè alunt.

Anglorū lapuuinga.

Sed hanc eſſe fulicam non patitur, quod Vergilius de fulica Georgicorum primo[1] *ad hunc modum ſcribit:*

Iam ſibi tum curuis malè temperat unda carinis,
Cùm medio celeres reuolant ex æquore mergi,
Clamoremq; ferunt ad littora, cumq; marinæ
In ſicco ludunt fulicæ.

Hinc ſatis liquet kyuuittam non eſſe fulicam, quum non ſit auis marina nec aquatica. Non deſunt qui fulicam gallinam illam nigram aquaticam, alba in fronte macula, eſſe uolunt. Sed iſti Vergil. et Ariſtotelis autoritate facilè erroris conuincuntur, quorum alter auem facit marinam, alter, nempe Ariſtoteles lib. octauo hiſtoriæ animalium, apud mare uiuere teſtatur. Quare quum paluſtris illa gallina neque auis ſit marina, neq; apud mare uictum petat, ſed in ſtagnis, paludibus, & recentibus aquis perpetuò degat: nec Vergilij fulica, nec Ariſtotelis κέπφος *eſſe poterit. Sed iam reſtat, ut quam auem fulicam eſſe iudicem, oſtendam.*

[p. 63]

Eſt auis marina, magnitudine monedulæ, ſed alis acutioribus & longioribus, colore tota albo, excepto nigro, quem in capite gerit cirro: roſtro etiam & pedibus puniceis. Hanc ego ſæpè in mari nauigans, ex eo

[1] Lib. I. l. 360—3.

Germany, who will have it that the Fulica is their Kywit, possibly resting their opinion on what they have read in Pliny of a tuft being attributed to the bird Fulica. The Kywit of the Germans is, however, smaller than a Cornix, with the plumage almost green and black on the whole back and head and neck, the belly white, a long and always upright feathery crest upon the head, and somewhat rounded wings, which during flight make a great hurtling, whence it is even named by foreigners Vannellus. It approaches waters for the sake of worms, on which alone it feeds, but does not enter them. It mostly lives in open country, and in places overgrown with heather. Our people often keep this bird in gardens, to destroy the worms.

And yet what Vergil in this manner writes in the first book of his Georgics of the Fulica will not permit this bird to be his Fulica:—

"And now the waters scarce restrain themselves from the ships' curving keels, while the swift Mergi wing their way once more out of the Ocean's midst, bringing their noisy voices to the shore, and while the Fulicæ, frequenters of the sea, disport themselves on land."

Hence it is clear enough that the Kywit is not the Fulica, since it is not a sea-bird nor a water-bird. There are not wanting those who would have that black Water Hen, with a white frontal patch, to be the Fulica[1]. But on the strength of Vergil and Aristotle such are easily convicted of mistake, for one of these makes it a sea-bird, and the other, namely Aristotle in the eighth book of his History of Animals, bears witness that it lives about the sea. Wherefore, since that Marsh Hen is neither a sea-bird nor seeks its food about the sea, but constantly haunts pools, and marshes, and fresh waters, it can neither be the Fulica of Vergil nor the *κέπφος* of Aristotle. But it still remains that I should shew what bird I judge the Fulica to be.

There is a sea-bird[2], like a Daw in size, but with the wings sharper and longer, wholly white in colour, save for a black patch which it bears on the head, and with the beak and feet of purplish red. I often, journeying upon the sea, have had

[1] *I.e.* the Coot (*Fulica atra*).

[2] The Black-headed Gull (*Larus ridibundus*).

tempore, quo hiſtoriam animalium Ariſtotelis legeram, conſideraui, tũ præſertim, quando uel deficiente uento, uel flante contrario, emiſſa anchora, uentum ſequundiorem quieſcentes expectauimus. Hæc ſtatim ſoluta anchora, gauijs comitata aduolat, ex purgamentis nauis eiectis, eſcæ nonnihil ſibi promittens, diutino clangore defatigata, tandem keph profert, ut lari cob. unde à noſtris marini cobbi dicuntur. Fieri poteſt, ut in fulicarum genere quædam ſint cinereæ, licet Plin. ubi ex autoritate Iubæ Diomedeas aues, fulicis ſimiles albas eſſe tradit, uideatur fulicas omnes albas facere, nam [p. 64] *non de eo, quod rarius, ſed frequentius accidit, in genere loquuntur claſſici ſcriptores. Nidulãtur lari & fulicæ in eiſdem locis, in excelſis nempe petris, & marinis rupibus.*

DE GAVIA.

Λάρος, *gauia, a ſe cob or a ſeegell.*

Gauiarum duo genera Ariſtot. facit: alterum album, quod apud mare, alterum cinerium, quod circa lacus & fluuios uictum quærit. Gauiam albam à fulica parùm differre arbitror, ſolo nimirum cirro et roſtro. Gauia cinerea, quæ ad flumina & lacus aſcendit, querula ſemper & clamoſa eſt. piſciculos captat & uermes ad ripas lacuum. huius generis eſt & alia parua auis, noſtrati lingua ſterna appellata, quæ marinis laris ita ſimilis eſt, ut ſola magnitudine & colore ab illis differre uideatur: eſt enim iſte larus, marinis minor & nigrior. Tota æſtate tam improbè clamoſa eſt, quo tempore parturit, ut iuxta lacus & paludes degentes, immodico clamore tantum nõ obtundat. hãc ego ſanè auem eſſe credo, cuius improba garrulitas adagio, Larus parturit, locum fecit. uolat ferè perpetuò ſuper lacus & paludes, nunquã quieſcens, ſed prædæ ſemper inhians. Nidulatur hæc in denſis arundinetis. Marinæ gauiæ in petris & rupibus maritimis nidificant.

this bird in mind, from the time that I read Aristotle's History of Animals, and then especially when through the wind failing or blowing contrary, the anchor being dropped, we have been calmly awaiting a more favourable wind. The anchor being weighed this bird immediately flies to us in the company of Gulls, promising something to itself by way of food out of the refuse cast forth from the ship; at last exhausted by its constant cries it merely utters "keph," as Gulls cry "cob." And hence they are called Sea-Cobs by our countrymen. It may be that some of the race of Fulicæ are grey, though Pliny, when on the authority of Juba he relates that the birds known as Diomede's are white like Fulicæ, seems to put down all Fulicæ as white; for classical authorities speak not in any class of what more rarely, but of what more frequently occurs. Gulls nest in the same places as do Fulicæ, forsooth on lofty crags and rocks about the sea.

Of the Gavia.

Λάρος, gavia, a se cob or a see-gell.

Aristotle makes two kinds of Gaviæ, one white, which seeks its food about the sea, the other grey, which seeks it round the lakes and rivers. Now I think that the white Gavia differs but little from the Fulica, only indeed as to the hood and beak. The grey Gavia, which comes up to our rivers and lakes, is always querulous and full of noise. It catches little fishes and eats worms upon the banks of lakes. There is another small bird of this kind, called Stern[1] in local dialect, which is so like the sea Lari that it seems to differ from them only in its size and colour; for it is a Larus, though smaller than the sea Lari and blacker. Throughout the whole of summer, at which time it breeds, it makes such an unconscionable noise that by its unrestrained clamour it almost deafens those who live near lakes and marshes. This I certainly believe to be the bird whose vile garrulity gave rise to the old Proverb "Larus parturit." It is almost always flying over lakes and swamps, never at rest, but always open-mouthed for prey. This bird nests in thick reed-beds. The sea Gaviæ breed on crags and rocks about the sea.

[1] The Black Tern (*Sterna nigra*).

[p. 65]

DE GALERITA.

Κόρυδος, ἡ κορυδαλός, *Ang. a lerk or a lauerock, Germa.* eyn lerch. *Plinius naturalis hiſtoriæ undecimo libro, galeritam Gallico uocabulo, poſtea alaudam eſſe dictam oſtendit. quare galeritæ potiùs uocabulo, quàm alaudæ Latinis utendum eſſe ex autoritate Plinij cenſeo.*

Aristoteles de galeritis [1].

Galeritarum duo ſunt genera: alterum terrenum criſtatum, alterũ gregale. nec ſingulare more alterius, uerùm colore ſimile, quãquam magnitudine minus, & galero carẽs, cibo uerò idoneũ, galeritæ nunquã in arbore conſiſtunt, ſed humi [2].

Prius hoc Ariſtotelis genus uariam habet in uarijs regionibus criſtam, alicubi ſemper apparentem: in alijs locis talem, ut pro arbitratu ſuo poſſit erigere aut deponere, quũ una eademq; ſit utriuſq; auis magnitudo. & galeritam hanc maiorem, Angli propriè lercam nominant. Alterum genus Aristotel. à noſtris fera alauda, à Germanis heid lerch *nominatur, in planis & locis erica conſitis, & ad ripas lacuũm, uermium cauſa, quibus uictitat, magna ex parte degit. Duplo ferè ſuperiore minus eſt, & roſtro tenui, & carne longè ſuauiſſima.*

a uuilde lerc or a heth lerk.

[p. 66]

Supereſt tertium galeritæ genus, Germanis copera, à lõgiſſima criſta, ut arbitror, ita dictum, Ariſtoteli planè incognitum: nam priùs Ariſtotelis genus eſſe non poteſt, quia minor eſt quàm ut illud eſſe poſſit: minùs autem illud genus eſſe non poteſt, quia galerum habet, qui Ariſtotelis poſteriori generi deeſt. Quare galerita iſta, Ariſtoteli fuit incognita. Et cùm Colonienſes aucupes coperam (quæ mediæ eſt magnitudinis inter Ariſtotelis

a uuodlerck.

[1] *Hist. An.* Bk IX. 101.
[2] *Hist. An.* Bk IX. 66.

OF THE GALERITA.

Κόρυδος, ἡ κορυδαλός, in English a lerk or a laverock, in German eyn lerch.

Pliny, in the eleventh book of his Natural History, has shown us that Galerita[1], taken from a Gaulish word, was called Alauda afterwards, wherefore I think that on Pliny's authority the name of Galerita should be used by those who write in Latin rather than Alauda.

ARISTOTLE, OF THE GALERITÆ.

Of Galeritæ there are two kinds, one is a crested ground-bird, but the other lives in flocks, not singly as the former. Yet in colour it is similar, though of a smaller size, and not having a crest. Moreover it is fit for food. The Galeritæ never sit upon a tree, but always on the ground.

The first kind given by Aristotle has in different lands a different crest, in one place always evident, elsewhere such that the bird can raise or lower it at will, although the size of either is one and the same. This larger Galerita Englishmen call the Lerc proper, while Aristotle's second sort is by our countrymen named a Wilde Lerc, and by the Germans a heid lerch; this for the most part lives in open country, and in places overgrown with heather, and on banks of lakes, for the sake of the worms on which it feeds. This bird is smaller by nearly one half than the aforesaid, with a slender beak, and flesh by far the sweetest.

There still remains a third kind of Galerita, the Copera of the Germans, thus named I believe from its very long crest, and certainly unknown to Aristotle, for it cannot be his first kind, inasmuch as it is smaller than that bird can be; likewise it cannot be the smaller sort, because it has a crest, which is not present in the latter kind. Wherefore this Galerita was unknown to Aristotle. And since the fowlers of Colonia [Cullen] with one accord assure us that the Copera (which is midway in size between Aristotle's crested Galerita

[1] Galerita is usually supposed to have some connexion with the Latin galea = a helmet.

galeritam criſtatam, & non criſtatam) concordibus adfirment ſuffragiȷs, hanc nullã habere peculiarem cantiunculam, ſed ineptè aliarum, quibuscum uictitat, auium uoces referre, adducor planè ut credam hanc eſſe recentiorum Græcorum corydon, cuius in ſequenti adagio mentio eſt ἐνάμουσις[1] καὶ κόρυδος φθέγγεται, *& in hoc uerſu:*

εἰ κύκνῳ δύναται κόρυδος παραπλήσιον ἄδειν.

Nã galerita maior, pulchrè & ſuauiter cantat, & minorem cantu non minùs ualere tradunt aucupes. Hæc igitur quum uoce nihil poſſit, ſed ineptè tantùm aliarum uoces, ſuo garritu mentiri, recentiorum Græcorum erit corydos.

[p. 67]

DE GALLIS ET gallinis.

Ἀλέκτωρ, *gallus, Anglicè a cok, Germanicè* eyn hän. Ἀλέκτορις, *gallina, Anglicè a hen, Germanicè,* eyn hen. *Saxones dicunt* eyn hön.

VARRO[2] DE RE RUSTICA LIBRO TERTIO.

Ruſticæ gallinę. Villaticę.

Gallinæ ruſticæ, ſunt in urbe raræ, nec ferè manſuetę ſine cauea uidentur Romæ, ſimiles facie non his uillaticis gallinis noſtris, ſed Africanis aſpectu, ac facie incontaminata. In ornatibus publicis ſolent poni cum pſiticis ac merulis albis, item id genus rebus inuſitatis. Neque ferè in uillis oua ac pullos faciunt, ſed in ſyluis.

Africanę. Meleagrides.

Gallinæ Africanæ, ſunt grandes, uariæ, gibberæ, quas Meleagrides Græci appellant. Hæ nouiſſimæ in triclinium ganearium introierunt, [p. 68] è culina propter faſtidium hominũ. Venerunt[3] propter penuriam magno. De tribus generibus, gallinæ ſaginantur maximè uillaticę. Eas in-

[1] Undoubtedly the reading should be:—ἐν ἀμούσοις = among those with little voice. Both these proverbs are to be found in the *Adages* of Erasmus, Chil. II. Cent. ii. 92.

[2] Bk III. cap. ix.

[3] No doubt a misprint for 'veneunt' = are sold.

and the non-crested) has no song of its own, but feebly imitates the notes of other birds with which it feeds, I am assuredly led to believe that it must be the Corydos of the later Greeks, of which mention is made in the proverb below:—

Ἐνάμουσις καὶ ὁ κόρυδος φθέγγεται,

and in this verse:—

εἰ κύκνῳ δύναται κόρυδος παραπλήσιον ᾄδειν.

For the larger Galerita sings fairly and sweetly, and the fowlers say that in its song the smaller kind is worth no less. Wherefore the third kind, since it has no power of voice except feebly to imitate the voices of the others by a twitter of its own, will be the Corydos of the later Greeks.

OF THE GALLI AND GALLINÆ.

Ἀλέκτωρ, gallus, in English a cok, in German eyn hän.

Ἀλέκτορις, gallina, in English a hen, in German eyn hen. The Saxons say eyn hön.

VARRO, IN HIS THIRD BOOK DE RE RUSTICA.

The wild Gallinæ are rare in a city, and are scarcely seen at Rome tame, unless in a cage: they are not in appearance like the Gallinæ of our country-houses, but in look recall the African, and have the face unmarked[1]. During public festivities these birds are wont to be exhibited with Psitaci, white Merulæ, and other unfamiliar kinds of that description. They do not usually lay their eggs or hatch their young at country-houses, but among the woods.

The African Gallinæ, which the Greeks call Meleagrides, are big, speckled, and hunch-backed. They have been the last to enter the dining room of eating-houses from the kitchen through people's fastidiousness. And from their rarity they have advanced to a great price. Of the three kinds those of the country-house are chiefly fattened. These they keep shut up in

[1] It is impossible to reconcile the statements of Varro and Columella as they stand. Various alterations of the text, which is possibly unsound, have been suggested.

cludunt in locum tepidum & anguſtum & tenebroſum, quòd motus earum & lux pinguedini inimica, electis ad hanc rem maximis gallinis, nec continuò his, quas Melicas appellant falſò, quòd antiqui ut thetim thelim dicebãt, ſic Medicã Melicã uocabant. Hæ primò Medicę dicebantur, quia ex Media propter magnitudinem erant allatæ.

Gallus medicus a bauncok or a cok of kynde.

Columella[1].

Gallinarũ alię ſunt cohortales, alię ruſticę, alię Africanę. Cohortalis eſt auis, quæ uulgò per omnes ferè uillas conſpicitur. Ruſtica, quæ non diſſimilis uillaticę, per aucupem decipitur, eaq; plurima eſt in inſula, quã nautæ in Liguſtico mari ſitam, producto nomine alitis, gallinariã uocitauerunt. Africana eſt, quam plerique Numidicam dicunt, meleagridi ſimilis, niſi quòd rutilam galeam, & criſtam capite gerit, quæ utraque in Meleagride ſunt cœrulea.

Cohortalis. Ruſtica. [p. 69] Africana.

Plinius[2].

Simili modo pugnant Meleagrides in Bœotia. Africæ, hoc eſt gallinarum genus gibberum, uarijs ſparſum plumis, quę nouiſſimę ſunt peregrinarum auium in menſis receptæ, propter ingratum uirus: uerùm Meleagri tumulus nobiles eas fecit.

Africæ.

Aristoteles[3].

[p. 70] Item Hadrianæ paruo quidem ſunt corpore, ſed quotidie pariunt. Ferociunt tamen & pullos ſæpè interimũt. Color his uarius. Oua[4]

[1] *De re rustica*, Lib. VIII. cap. ii.
[2] *Hist. Nat.* Lib. X. cap. xxvi.
[3] *Hist. An.* Bk VI. 1.
[4] *Hist. An.* Bk VI. 5.

a warm, narrow, and dark place, for exercise and light hinder the fattening. The largest birds are chosen for this purpose, and not always those which men mistakenly call Melicæ, because the ancients, as they used to say Thelis for Thetis, also used to call Medica Melica. At first they were called Medicæ because on account of their size they were brought hither out of Media.

COLUMELLA.

Of Gallinæ some are court-yard birds, others again are wild, others are African. The court-yard bird is that which commonly is seen at nearly every country-house. The wild sort, which is not unlike that of the country-house, is trapped by bird-catchers. It is abundant in the island lying in the Ligurian sea, which sailors, lengthening the bird's name out, have called continuously Gallinaria. The kind from Africa, which many call Numidica, is like the Meleagris, save that on its head it bears a helmet and a crest of red, but in the Meleagris both of these are blue[1].

PLINY.

In a like way the Meleagrides fight in Bœotia. The Africæ, that is a hunch-backed kind of Gallinæ, are sprinkled here and there with variegated feathers; and they are the last of foreign birds to be received at table, on account of their unpleasant flavour: but the tomb of Meleager has ennobled them[2].

ARISTOTLE.

Likewise the Hadrianic birds are small indeed in body, but they lay their eggs daily. Yet they are fierce and often kill their chicks. They are of varied

[1] See Art. Guinea Fowl in Prof. Newton's *Dict. B.* p. 399.

[2] The reader may here be referred to any work relating to mythology.

alia candida ſunt, ut columbarum & perdicum, alia pallida, ut paluſtrium, alia punctis diſtincta, ut Meleagridum & phaſianorum.

In paucis iſtis, quos recenſui autoribus, nõ paucæ ſunt de rebus, quas tractauêre, cõtrouerſiæ. Varro primũ gallinas ruſticas non uillicatis[1] *ſed Africanis ſimiles eſſe ſcribit. Columella autẽ ruſticam non diſſimilem eſſe uillaticæ tradit. Varro Africanas, meleagrides facit, quod & Plinius etiam facere uidetur. Columella autem uarijs notis Africanas à meleagridibus diſtinguit. Ariſtoteles Hadrianas gallinas facit uarias, ut Plinius itidem facit, & paruo corpore. Varro Africanas, quas non alias eſſe conſtat quàm Hadrianas, uarias & grandes facit. Verùm tanta autorum inter ſe diſſidia componere, penes me non eſt. Sed quid de generibus iſtis ſentiam, paucis aperiam. Gallina apud nos ruſtica nuſquam reperitur, ſi gallina illa, quam morhennam uocant non ſit, quam uarijs de cauſis* [p. 71] *antè attagenem eſſe conieci. Columellæ meleagrides uidentur illæ eſſe aues, quas nonnulli pauones Indicos appellant: nam illas paleis*[2] *& criſtis cœruleis eſſe, in confeſſo eſt.*

a kok of inde.

DE GALLINAGINE.

Ασκαλώπαξ, *gallinago, Anglicè a uuod cok, Germanicè* eyn holtz ſnepff.

Aristoteles[3].

Gallinago per ſepes[4] hortorum capitur, magnitudine quanta gallina eſt, roſtro longo, colore attagenæ, currit celeriter, & hominem mirè diligit. Hæc in arbore nunquam[5] ſedet, & humi nidulatur.

[1] A misprint for 'uillaticis.'

[2] Perhaps the reading should be galeis, cf. p. 69, ll. 6—7.

[3] *Hist. An.* Bk IX. 102. 66.

[4] Aristotle has in addition—ἕρκεσι = in nets.

[5] *Hist. An.* Bk IX. 66.

colouring. Of certain kinds of birds the eggs are white, as those of Columbæ and Perdices, others are pale, as those of marsh-birds; others marked with spots, as those of Meleagrides and Phasiani.

In those few authors, whose works I have scanned, not a few of the things which they have treated are disputable. First Varro tells us that the wild-bred Gallinæ are not like those of country-houses, but the African; while Columella states that the wild sort is not unlike that of the country-house. Varro makes the Africanæ to be the Meleagrides, which Pliny also seems to do. Yet Columella separates the Africanæ from the Meleagrides by various characters. And Aristotle makes his Hadrianic fowls of various colours, as does Pliny also, and of little size. But Varro makes the Africanæ big and of varied colours, though it is quite clear that they are nothing but the Hadrianic birds. But after all it is not in my power to adjust the mutual differences of authors, when so great; and yet in a few words I will disclose what I think of these kinds. The wild Gallina is not found with us in any part, if it be not that which they name Morhen, and this I formerly conjectured to be the Attagen for several reasons. The Meleagrides of Columella seem to be those birds which some call Indian peacocks[1], for they are admitted to have wattles and blue crests.

Of the Gallinago.

Ἀσκαλώπαξ[2], gallinago, in English a wod cok, in German eyn holtz snepff.

Aristotle.

The Gallinago is taken among the hedges of our gardens; it is of the size of a Gallina, but has a long bill, and the colour of the Attagena: it runs with speed, while it is wonderfully fond of man. This bird never sits on a tree and it nests on the ground.

[1] Turner was, of course, wrong in his conjecture.

[2] Turner makes σκολόπαξ the same as ἀσκαλώπαξ.

Gallinagines apud noſtrates nunquam, niſi hyeme uidentur, quare de prole & modo nidulandi, nihil habeo, quod dicã. Capitur apud Anglos diluculo potiſſimùm & crepuſculo in ſyluis, retibus in loco arboribus uacuo, ſuſpenſis, & ueniente aue demiſſis.

De colio, sive galgulo[1], ut vertit Gaza, ex Aristotele.

[p. 72] Galgulo magnitudo quanta ferè turturi eſt: color luteus, lignipeta hic admodum eſt, magna'que ex parte macerie[2] paſcitur, uocẽ emittit grandem, incola maximè Peloponeſi hęc auis eſt.

Omnia, quæ Ariſtoteles hactenus colio, ſiue galgulo tribuit, Anglorum huhulo, & Germanorum grunſpechto (ſi incolam maximè eſſe Peloponeſi exceperis) conueniunt. Nam turturem ferè magnitudine æquat, lignipeta eſt: maceriem contundit, & uocem grandem emittit. Sed nihil hic definio, ſed inquiro tantùm. Galgulus Plinio icteros Græcè dicitur, & Ariſtoteli, ſi Theodoro fidimus, etiam κελεὸς. *Quanquam mihi textum Græcum conſulenti, alia auis* κολιός, *& alia* κελεός *uidetur: nam* κολιός ἐστὶ ξυλοκόπος σφόδρα, καὶ νέμεται ἐπὶ τῶν ξύλων τὰ πολλά. *Id eſt, colius eſt lignipeta ualde, & magna ex parte ad ligna paſcitur,* ὁ μὲν γὰρ κελεός παρὰ ποταμόν οἰκεῖ καὶ λόλμας[3], *quæ uerba Theodorus circa fruteta & nemora reddidit ſed recte'ne an ſecus, doctis iudicandum relinquo. Vidi in Alpibus abieti inſidentem auem, magnitudine turturis, uiridibus ueluti maculis in luteo diſtinctam, quæ tota corporis effigie* [p. 73] *picũ Martium retulit, ſed caput reliquo corpori (ſecus*

[1] *Hist. An.* Bk VIII. 44.

[2] Possibly a misprint for 'materie.'

[3] *Hist. An.* Bk IX. 22. In his errata Turner alters λόλμας to χλόχμας, but evidently he means λόχμας.

Woodcocks are never seen with us save in the winter, wherefore I have naught to say about their young or mode of nesting. They are chiefly caught in England in the woods at daybreak and at dusk, by means of nets hung in some place devoid of trees, and dropped when the bird comes.

Of the Colius, or Galgulus, as Gaza renders it, from Aristotle.

Of the Galgulus the size is almost that of the Turtur: it is yellowish in colour, and hacks timber very much, and for the most part feeds on trees: it utters a loud cry. This bird is mainly an inhabitant of the Peloponnese.

All that Aristotle has so far attributed to the Colius or Galgulus is in agreement with the Huhol of the English and the Grunspecht of the Germans (if one may except its being chiefly an inhabitant of the Peloponnese). For it is nearly equal to the Turtle-Dove in size; it hacks the timber, hammers rotten wood, and utters a loud cry. But I give no decision here, I only ask. The Galgulus of Pliny is said to be called the Icteros in Greek, and if we trust to Theodorus [Gaza] is also the *κελεὸς* of Aristotle. Though, on consulting the Greek text *κολιὸς* seemed to be one bird, and *κελεὸς*[1] another, for the reading was:—*κολιός ἐστι ξυλοκόπος σφόδρα, καὶ νέμεται ἐπὶ τῶν ξύλων τὰ πολλά.* That is, the Colius is especially a wood-hunter and for the most part feeds on wood, *ὁ μὲν γὰρ κελεὸς παρὰ ποταμὸν οἰκεῖ καὶ λόχμας*, which words Theodorus renders "around the thickets and the groves," but whether rightly so or otherwise I leave to be decided by the learned. In the Alps I saw sitting upon a fir a bird of the size of a Turtle-Dove, marked as it seemed with green patches on yellow, which to me in the whole aspect of the body called to mind the Picus Martius, save that its head was like in colour to the rest of its body

[1] Turner appears to have had a text with the word *κελεὸς* in one place instead of *κολιός*.

atq; in pico ſit) colore fuit ſimile, tibijs fuit breuibus, et capite erecto, & roſtro longiuſculo. An hæc galguli ſpecies fuerit, nihil ſtatuo, ſed fuiſſe ſuſpicor.

DE GRACVLIS.

Eraſmus in eruditiſſimo adagiorum opere, quoties κολοιὸς *occurrit (occurrit aūt non rarò) graculum reddit. Theodorum Gazam hac in re, licet aliâs libenter, minimè ſecutus, qui* κόλοιον *ſemper monedulam uertit. Ego quoq; hac in re Eraſmum potiùs quàm Gazam, uarijs de cauſis, imitari decreui:*

ARISTOTELES SECUNDÙM TRANSLATIONEM GAZÆ[1].

Monedularum tria ſunt genera: unum, quod graculus uocatur, magnitudine quanta cornix, roſtro rotundo, rutilo. Alterum, lupus cognominatum, paruū & ſcurra. Tertium, quod familiare, Lydiæ ac Phrygiæ terræ, idemq; palmipes eſt.

[p. 74] *Primum graculorum genus, quod Græci* κορακίαν *uocant, Plinio Pyrrhocorax eſt, Anglis a cornish choghe, Germanis* eyn bergdöl, *cornice paulò minor eſt, roſtro luteo, paruo, & in fine nonnihil adunco, frequens eſt in alpibus, & apud Anglos in Cornubia, uocem habet monedula acutiorem, & magìs querulam. Secundum genus* λύκος καὶ βωμολόχος, *Græcè dictum, Latinis propriè monedula, quaſi monetula à moneta dicitur, quam ſola auium, ut inquit Plinius, furatur. Aurum non omnia tria genera furantur, ſed ſecundum genus tantùm, quare*

Monedula. Plin. li. 10.

[1] *Hist. An.* Bk IX. 100.

(otherwise than it is in Picus), and the legs were short, the head was upright, the beak rather long. As to whether this may have been a kind of Galgulus, I do not certify, but I suspect it to have been.

Of the Graculi.

Erasmus in his very learned work on Proverbs, as often as *κολοιὸς* occurs (and it occurs not seldom) renders it by Graculus, in this thing following by no means Theodorus Gaza—though at other times he does so freely—who in every case renders *κολοιὸς* by Monedula. And in this thing I also have determined for divers reasons here to imitate Erasmus rather than Gaza.

Aristotle according to the translation of Gaza.

Of Monedulæ there are three sorts: the first, which is called Graculus, in size as big as Cornix with a curved red bill. The next, also named Lupus, small, and a mimic. The third, which is well known in Lydia and Phrygia, is web-footed.

Now the first kind of Graculi, which the Greeks call *κορακίας*, is the Pyrrhocorax of Pliny and the Cornish Choghe of Englishmen, eyn bergdöl of the Germans. It is a little smaller than the Cornix, with a yellow bill[1], not large, and somewhat hooked towards the tip, it is abundant in the Alps and in Cornwall in England. It has a sharper and more querulous cry than the Monedula. The second sort called *λύκος* and *βωμολόχος* in Greek, is by the Latins strictly named Monedula, as if it were Monetula, from the Moneta [money] which alone of birds, as Pliny says, it steals. The three kinds do not all steal gold—only the second does—

[1] Here there is an evident confusion between the Chough (*Pyrrhocorax graculus*) with its red bill, and the yellow-billed Alpine Chough (*P. alpinus*).

ſecundum genus ſolum erit monedula, de cuius furacitate pulchrè etiã his uerſibus ſcribit Ouidius[1]*:*

Mutata eſt in auem, quæ nunc quoq; diligit aurum,
Nigra pedes, nigris uelata monedula pennis.

Monedulã Angli uocãt, a caddo, a chogh or, a ka. Germ. eyn döl, *& Saxon.* eyn älke. *Multò minor eſt pyrrhocorace monedula, & in ſyluis nidulatur, & in cauis arboribus, & in templorũ turribus. Tertium genus, Ariſtoteles* 8 *libro hiſtoriæ animalium*[2] *ita deſcribit:* Palmipedum grauiores circa lacus & amnes uerſantur, ut anas, phalaris, urinatrix. Ad hęc [p. 75] boſca ſimilis anati, ſed minor, & qui coruus appellatus eſt, cui magnitudo quanta ciconiæ, ſed crura breuiora, palmipes natansq; eſt, colore niger, inſidet arboribus & nidulatur in ijs. Hęc Ariſtoteles. *Coruus iſte, niſi fallar, Plinio phalacrocorax eſt, & Heluetiorum Vualtrapus, de quo Plin. ad hunc modum ſcribit*[3]*:* Iam & in Gallia Hiſpaniaq; capitur attagen, & per alpes etiam, ubi & phalacrocoraces, Balearium inſularum peculiares: ſicut & alpium pyrrhocorax. *Et alibi de eodem*[4]*:* Quædã animalium naturaliter caluent, ſicut ſtruthiocameli, & corui aquatici, quibus apud Græcos nomen eſt inde.

Phalocrocorax.

Phalocrocorax .i. coruus caluus.

[1] *Metamorph.* Lib. VII. ll. 467—8.
[2] Bk VIII. 48.
[3] *Hist. Nat.* Lib. X. cap. xlviii.
[4] *Op. cit.* XI. cap. xxxvii.

wherefore this second kind alone shall be Monedula; moreover Ovid happily describes its thievish habits in the following lines:—

> Was changed into a bird, which even now loves gold,
> Monedula the black of foot, in plumage black arrayed.

The English call the Monedula a Caddo, Chogh, or Ka; Germans eyn döl; and Saxons eyn älke. The Monedula is much smaller than the Pyrrhocorax, and nests in woods and hollow trees and towers of churches. The third kind is thus described by Aristotle in the eighth book of his History of Animals:—

Of web-footed birds the heavier haunt lakes and rivers, as the Anas, Phalaris, and Urinatrix. Add to these the Bosca, which is like the Anas but smaller, and that which is called Corvus, whose size is that of a Ciconia, but it has shorter legs; it is web-footed and a swimmer: black in colour, it perches on trees, and nests in them. So far Aristotle.

Unless I err, this Corvus is the Phalacrocorax of Pliny and the Swiss Waltrapus[1], of which Pliny writes after this fashion:—

Further, the Attagen is caught in Gaul and Spain, and even on the Alps, where Phalacrocoraces also are, proper to the Balearic isles, as the Pyrrhocorax is to the Alps.

And in another place of the same bird:—

Some animals are naturally bald, as Struthiocameli and Corvi Aquatici, whence is their name among the Greeks.

[1] Mr Rothschild identifies this bird with *Comatibis comata* = *C. eremita* (L.), no doubt rightly. See *Bull. Brit. Orn. Club*, XII. p. 56. *Novitates Zoologicæ*, 1897, p. 371, and Pliny, Lib. X. cap. xlviii.

Iam ut ſciatis qualis'nam auis ſit Heluetiorum Vualtrapus, quam conijcio phalacrocoracem eſſe, & tertium genus graculi, auis eſt corpore longo, & ciconia paulò minore, cruribus breuibus, ſed craſſis, roſtro [p. 76] *rutilo, parùm adunco, & ſex pollices longo, albam quoque in capite maculam, & eàm nudam, niſi malè memini, habuit. Si palmipes ſit, & interdum natet, indubitanter tertium graculorum genus eſſe adfirmarem: uerùm licet auem in manibus habuerim, an palmipes fuerit nec'ne, & caluus, non memini: quare donec iſthæc certiùs nouero, nihil ſtatuam.*

Graculus nucifrag⁹ eyn nouſbrecher.

Præter hæc tria graculorum genera ab Ariſtotele deſcripta, noui & quartum genus, quod in alpibus Rheticis uidi, Ariſtotelis lupo minus, nigrum & albis maculis per totum corpus, more ſturni diſtinctum, garrulitate ſuperiora genera multùm ſuperans, ſemper in ſyluis & montibus degens: cui Rheti nucifragæ nomen, à nucibus quas roſtro frangit & comedit, indiderunt.

DE GRVE.

Γερανός, *grus, Anglicè a crane, Germanicè* eyn krän / oder eyn kränich.

Aristoteles[1].

Alia de ultimis propè ueniunt, ut grues faciunt, quæ Scythicis ad paludes Aegypto[2], unde Nilus profluit, ueniunt: quo in loco pug- [p. 77] nare cũ pygmeis dicuntur. Non enim id fabula eſt, ſed certè genus tum hominum tum etiam equorum puſillum, ut dicitur eſt, deguntq; in cauernis, unde nomen troglodytæ, à ſubeundis cauernis accepêre. Grues[3] etiam multa prudenter faciunt: loca enim longinqua petunt, fui commodi gratia, & altè uolant, ut procul proſpi-

[1] *Hist. An.* Bk VIII. 75—76.
[2] A variant reading is τὰ ἄνω τῆς Αἰγύπτου.
[3] *Hist. An.* Bk IX. 70.

And now, that you may know what sort of bird the Switzers' Waltrapus may be, which I conjecture is the Phalacrocorax, and the third kind of Graculus, it is a bird long in the body, which is rather less than that of the Ciconia, and the legs short but stout, the bill reddish, a little hooked, and six inches in length—further it had a white spot on the head, and that, unless my memory fails me, bare. If it be web-footed and swim at times, I should affirm that it undoubtedly was the third kind of Graculus; but, though I have myself had the bird in my hands, I do not now remember whether it was web-footed or not, nor whether it was bald. Wherefore I will determine nothing, until I shall have a surer knowledge of these things.

Besides the said three kinds of Graculi described by Aristotle I know a fourth, which I have seen upon the Rhætic Alps, smaller than Aristotle's Lupus, black and marked with spots of white on the whole body, as a Starling is; it far surpasses all the above-named kinds in chattering; it always lives in woods and mountains. Now to this the Rhetians have given the name of Nucifraga, from the nuts which it breaks with its bill and eats.

OF THE GRUS.

Γερανός, grus, in English a crane, in German eyn krän, or eyn kränich.

ARISTOTLE.

Others come almost from earth's utmost parts, as do the Grues, which come from the Scythians to the Egyptian marshes, whence the Nile flows forth: in which place they are said to fight with Pygmies. And this is no mere fable, but assuredly there is, as it is said, a dwarf race both of men and horses, and they live in caves, whence they have got the name of Troglodytæ, from dwelling in caves. The Grues furthermore do many things with prudence, for they seek for their convenience distant places, and fly high that they may look out far, and, if they shall

cere poſſint, & ſi nubes tempeſtatem'ue uiderint, conferunt ſe in terrã, & humi quieſcũt. Ducem etiam habent, & eos, qui clament, diſpoſiti[1] in extremo agmine, ut uox percipi poſſit. Cùm conſiſtunt, cæteri dormiunt, capite ſubter alam condito, alternis pedibus inſiſtentes. Dux detecto capite, proſpicit, & quod ſenſerit, uoce ſignificat.

Pipers. *Vipiones Plin. dicuntur minores grues & iuniores,* [p. 78] *ut pipiones iuniores dicuntur columbæ. Apud Anglos etiam nidulantur grues in locis paluſtribus, & earum pipiones ſæpiſſimè uidi, quod quidam extra Angliam nati, falſum eſſe contendunt.*

DE HIRVNDINE.

Χελιδών, *hirundo, Anglicè à ſuuallouue, Germanicè* eyn ſchwalb. *Saxonibus eſt* eyn ſwale.

ARISTOTELES[2].

Hirundo carnibus ueſcitur, bis in anno parit, & tota hyeme latet. Omnino ratio brutorum, magnã refert uitæ humanę ſimilitudinem magisq; in minori genere, quàm in maiore. uideris intelligentiæ rationem, quod primum in auium genere hirundo in effingendo cõſtituendoq; nido oſtendit, confingit implicito luto, feſtucis ad normam lutariæ paleationis, & ſi quãdo luti inopia eſt, ſe ipſa madefaciens, uolutat in pul- [p. 79] uerem omnibus pennis. Stragulum etiam facit more hominum duriore primùm ſubijciens, & modicè totum conſternens, pro ſui corporis magnitudine.

[1] 'dispositi' is here apparently attracted to 'qui.'

[2] Four passages are incorporated in Turner's selection:—*Hist. An.* Bk VIII. 39, Bk VI. 36, Bk VIII. 107 and Bk IX. 51, 52.

have seen clouds or a storm, betake themselves to earth, and take rest on the ground. They have a leader also and those who, disposed at each end of the band, may call out, that their voice may be perceived. The others sleep when they alight, with the head hidden underneath the wing, standing alternately on either foot. The leader gazes round him with uncovered head, and by his cry gives notice of whatever he perceives.

The smaller, that is younger, Cranes are called by Pliny Vipiones, as young Doves are known as Pipiones. Cranes, moreover, breed in England in marshy places, I myself have very often seen their pipers[1], though some people born away from England urge that this is false.

OF THE HIRUNDO.

Χελιδὼν, hirundo, in English a swallowe, in German eyn schwalb. Among the Saxons it is eyn swale.

ARISTOTLE.

The Hirundo feeds on flesh, and lays eggs twice a year, for the whole winter also it lies hid. The way of brutes upon the whole bears a marked likeness to the life of man, and more so in the smaller than the larger kinds. One may observe the understanding way which the Hirundo foremost in the ranks of birds shews in the constitution and construction of its nest. It builds it by applying mud to straws, after the rule of daub and wattle work, and if there ever be a scarcity of mud, it wets itself and rolls itself in dust with all its feathers. It moreover makes a bedding after the manner of men, first laying a foundation of the harder stuff below, and moderately covering the whole in proportion to its size.

[1] Young pigeons are still called Pipers in England.

PLINIUS[1].

Hirundines luto conſtruunt, ſtramento roborant. Si quando inopia eſt luti, madefactum multa aqua pennis puluerem ſpargunt. Ipſum uerò nidum mollibus plumis flocciſq; conſternunt, tepefaciendis ouis, ſimul ne durus ſit infantibus pullis.

Hirundinum primũ genus.

Alterum genus eſt hirundinum ruſticarum & agreſtiũ, quæ raro in domibus, diuerſos figura, ſed eadẽ materia nidos confingunt, totos ſupinos, faucibus porrectis in anguſtum, utero capaci: mirum qua peritia occultandis habiles pullis, & ſubſternendis molles.

Secundũ. [p. 80]

Tertium hirundinum genus eſt, quæ ripas excauant, atque ita † internidificant. Non faciunt hæ nidos, migrãtq; multis diebus antè, ſi futurum eſt, ut auctus amnis attingat.

Tertium. † in terra fœtificãt.

DE APODIBUS, EX ARISTOTELE[2].

Nonnullæ aues depedes[3] ſunt, quæ ob eam rem apodes à paruitate pedum nuncupantur. quod genus auiculæ, pennis plurimùm ualet, ſicut & cætera quoque propè ſimilia, ut pennis præualere, ſic pedibus degenerare uidentur. ut hirundo & falcula ſiue riparia[4]. Hæc enim omnia, & moribus, & uolatu, & ſpecie proxima inter ſe conſpiciuntur. Apparet apes omnibus anni temporibus: riparia æſtate tantùm cùm imber inceſſit: tum enim & apparet & capitur.

[p. 81] Riparia ſiue falcula

1 *Hist. Nat.* Lib. x. cap. xxxiii.
2 *Hist. An.* Bk I. 10.
3 κακόποδες in the Greek.
4 These two words are probably interpolated.

PLINY.

Hirundines build nests of mud, and strengthen them with straw. And if there ever be a scarcity of mud, they sprinkle a good store of water from their feathers on the dust, which is thus moistened. The nest itself they further line throughout with soft feathers and wool, to thus keep the eggs warm, and also that it may not be too hard for the young chicks.

There is another sort of the Hirundines of the country and the fields, which rarely build their nests in houses, different in shape, but of the same material, and facing wholly upwards, having entrances prolonged into a strait with a capacious belly[1]; it is wonderful how skilfully they are adapted for concealing young, and soft for them to lie upon.

There is a third kind of Hirundines which bore in banks, and thus breed within holes. These make no nests, and migrate many days before, if it be likely that the stream in flood should reach them.

OF THE APODES, FROM ARISTOTLE.

Some birds are weak-footed, and for that reason from the smallness of their feet are known as Apodes. This kind of little bird is very strong upon the wing, just as some others that are nearly like it seem to lose in strength of foot proportionately as they gain in power of flight, as the Hirundo and the Falcula, in other words Riparia. For all these in their habits, flight, and look seem very near each other. The Apes may be seen at all times of the year, but the Riparia only in summer, when the rains begin: for then it is both noticed and is caught. In

[1] Pliny evidently refers to *Hirundo rufula*, which builds a flask-shaped nest against a cliff.

Apodes. deniq; rara hæc auicula eſt. Apodes[1], quos aliq cypſellos uocant, ſimiles eſſe hirũdinum, iam dictum eſt: haud enim ab hirundine diſcerni poſſunt, niſi quòd tibijs ſunt hirſutis. Nidum ſpecie ciſtellæ[2] productæ lõgiùs fictæ ex luto, imò aditu dato arctiſſimo faciunt, idq; locis anguſtis[3], intrà ſaxa & ſpecus, ut & belluas, & homines poſſint deuitare.

PLINIUS[4].

Apodes. Plurimũ uolant, quę apodes uocãtur, quia carẽt uſu pedum. Ab alijs Cypſelli appellantur, hirundinum ſpecie. Nidificant in ſcopulis. Hæ [p. 82] ſunt, quæ toto mari cernuntur: nec unquam tam longo naues, tamq; continuo curſu, recedunt à terra, ut non circumuolitẽt eas apodes. Cætera genera reſidunt, & inſiſtunt: his quies niſi in nidis nulla: aut pendent, aut iacent. *Hactenus Plinius & Ariſtoteles.*

Ariſtoteles tria tantùm hirundinum genera facit: domeſticas, apodes, & falculas. Plinius autem quatuor genera facere uidetur: domeſticas, ruſticas, apodes & riparias. Quod ſi uerum ſit, hirundines domeſticæ, ſanguinolento pectore nobiles, erunt primum genus. Secundum genus maximæ illæ & nigerrimæ hirundines gregatim plerumq; uolantes, facere uidentur. Tertium genus, hirundines quæ in ſummis turribus & altis templorum feneſtris nidulantur, efficiunt. Quartum genus ripariæ ſiue falculæ erunt. Quòd ſi iſta diuiſio parùm arrideat, ad primum genus referantur Hirũdines domeſticę. *hirundines illæ in domibus ruſticorum ſemper nidificantes, quæ à reliquis generibus, duæ ſanguinolentæ*

[1] *Hist. An.* Bk IX. 108.
[2] κυψέλεσιν is the word in Greek.
[3] στενῷ; another reading is στεγνῷ = under cover.
[4] *Hist. Nat.* Lib. X. cap. xxxix.

short this little bird is rare. The Apodes, which some call Cypselli, are like Hirundines, as has been said before, for they are not to be distinguished from the Hirundo, save by having hairy legs. The nest which they construct looks like a little basket formed of mud somewhat drawn out, an entrance of the straitest opening beneath; and this they place in cracks within the rocks and caves, that they may avoid both beasts and men.

PLINY.

The birds which, because they cannot use their feet, are called Apodes, live chiefly on the wing. They are named Cypselli by some, in aspect they are like Hirundines. They nest in crags. These are they which are seen on all parts of the sea, nor do ships ever leave the land upon so long or so continuous a course but that the Apodes still fly around them. The other kinds alight and settle, but for these there is no rest save in their nests alone, they either hang or lie. So far Pliny and Aristotle.

Now Aristotle makes only three kinds of Hirundines, those of the house, the Apodes and the Falculæ. Yet Pliny seems to make four kinds, those of the house, the Rusticæ, the Apodes, and the Ripariæ. If that be true, our House Swallows, well known from their blood-coloured breast, will be the first-named kind. Those very large and black Swallows, that mostly fly in flocks, appear to form the second kind. Those Swallows which make nests upon the tops of towers, and in lofty church windows, constitute the third. And the Ripariæ or Falculæ will be the fourth. But should the said division not approve itself, then to the first-named kind may be referred those Swallows which invariably build on the houses of the country people. Two patches of a blood colour, which one may see on each side

maculæ, quas utrinque in pectore uideas, diſtinguunt, quod Ouidius[1] *his uerſibus pulchrè oſtendit:*

[p. 83] *Altera tecta ſubit, nec adhuc de pectore cædis*
Exceſſêre notæ, ſignataq; ſanguine pluma eſt.

Hoc primum genus Angli a ſuuallouu nominãt & Germani eyn ſchwalb.

Apodes. *Secundum genus faciunt apodes tam maiores quàm minores. maiores uoco maximas illas hirundines, gregatim & altiùs cæteris uolantes, quæ in arbore, more hirundinum aliarum nunquam conſiſtere uiſuntur. minores uoco, quæ in ſcopulis, templorum feneſtris æditioribus & ſummis turribus nidos figunt. Maiores Germani uocant* geyr ſwalben, *Angli the great ſuuallouues. Minores Angli uocant rok martinettes or chirche martnettes, Germani uocant* kirch ſwalben.

Falcula ſiue riparia *Tertium genus, quod in ripis nidulatur, Angli a bank martuet*[2], *Germani* eyn über[3] ſwalbe, *aut* ſpeiren *nominant.*

De hæmatopodibus, ex Plinio[4].

Roſtrum & prælonga crura rubra hæmotopodi[5] ſunt, multò Porphyrione minori: quãquam [p. 84] eadem crurum altitudine. Naſcitur in Aegypto. Inſiſtit ternis digitis, præcipuum ei pabulum muſcæ. Vita in Italia paucis diebus.

Eſt apud Anglos in locis paluſtribus auis quædam longis & rubris cruribus, noſtra lingua redſhanca dicta, cui an deſcriptio hæmotopodis Pliniani conueniat necne, qui apud Anglos degunt, inueſtigent & examinent.

DE IVNCONE.

Σχοίνικλος, *iunco, Anglicè a rede ſparrouu, Germanicè* eyn reydt müß. *Iunco, ut ſcribit Ariſtoteles octauo hiſtoriæ animalium, & capite tertio, ad ripas lacuum & fluminum uictitat, & caudam frequenter motitat, & ex eodem conſtat, auem eſſe paruam: nam turdo minorem*

[1] *Metam.* Lib. VI. ll. 669—70.
[2] A misprint for 'martnet.'
[3] A misprint for 'ůfer.'
[4] *Hist. Nat.* Lib. X. cap. xlvii.
[5] Another reading is 'Himantopus,' but the Stilt-Plover has not a red bill.

of the breast, distinguish these from the remaining sorts, as Ovid prettily sets forth in these verses:—

"The other haunts our roofs, nor have the marks of slaughter yet departed from its breast, and its plumage is stained with blood."

The English call this first kind a Swallow, and the Germans eyn schwalb.

The Apodes, the greater and the less, compose the second kind. I call greater those very great Swallows that fly in flocks, and higher than the rest, which are never observed to settle on a tree, after the manner of our other Swallows. I call less those which fix their nests to rocks, lofty church windows and the tops of towers. The greater kind the Germans call geyr swalben, and the English the Great Swallowes; but the less the English call rok martinettes or chirche martnettes, the Germans kirch swalben.

The third kind, that which breeds in banks, the English name a bank martnet, the Germans eyn ůfer swalbe or speiren.

OF THE HÆMATOPODES, FROM PLINY.

The Hæmatopus has its bill and very long legs red, and is much less than the Porphyrio, though of the same height of leg. It is native in Egypt. It stands on three toes to a foot; flies are its favourite food. It lives in Italy but a few days.

There is in marshy places in England a certain bird with long red legs, called Redshank in our tongue, but whether the description of the Hæmatopus of Pliny agrees with this or not let those who live in England seek out and enquire.

OF THE JUNCO.

Σχοίνικλος, junco, in English a rede sparrow, in German eyn reydt můss.

The Junco, as Aristotle writes in the eighth book of his History of Animals, and in the third chapter, lives on the banks of lakes and streams, and flirts its tail continually; and it is clear from him that it is a small bird, for he makes

facit. Ego igitur quum nullam aliam nouerim auiculam, iuncis & harundinibus inſidentem, præter Anglorum paſſerem harũdinarium, illum iunconem eſſe iudico. Auis eſt parua, paſſere paulò minor, cauda longiuſcula & capite nigro. cætera fuſca.

De lingulaca, ex Aristotele[1].

Lingulaca, quæ Græcè γλώττις dicitur lin- [p. 85] guam exerit longam, unde nomẽ habet, una eſt è coturnicum ducibus, formam habet auium lacuſtrium.

De lagopode ex Plinio[2].

Præcipuo ſapore lagopus eſt: pedes leporino uillo ei nomen hoc dedêre. Cętero candidę[3], columborum magnitudine, non extra terram[4], in qua naſcitur, eam ueſci: quando nec uiua manſueſcit, & corpus occiſę, ſtatim marceſcit. Eſt & alia, nomine eodem, à coturnicibus magnitudine tantùm differens, croceo tinctu cibis †aptiſſima. *Huius hoc uerſu Martialis*[5] *meminit:*

† gratiſſima.

Si meus aurita gaudet lagopede Flaccus.

DE LIGVRINO SIVE SPINO.

Ακανθίς, *ſpinus, ſiue ligurinus, Anglicè, a grene finche, ut coniicio, Germanicè,* eyn firſfincke.

Aristoteles[6].

[p. 86] Ligurini, & uita & colore ignobiles ſunt, ſed ualent uocis amœnitate[7], & ex auium albo ſunt, quę carduorum ſemine ueſcuntur[8]. Florus, ſpinus, & ægithus, odium inter ſe exercẽt. Spinus etiam bellum cum aſino gerit.

[1] *Hist. An.* Bk VIII. 83.
[2] *Hist. Nat.* Lib. X. cap. xlviii.
[3] After 'dedere' substitute a comma for the full stop.
[4] Pliny seems to have written 'facile' here, in addition.
[5] *Epigr.* Lib. VII. lxxxvi.
[6] Three passages are here combined:—*Hist. An.* VIII. 42, IX. 22, IX. 92.
[7] The words of Aristotle are φωνὴν μέντοι λιγυρὰν ἔχουσιν.
[8] See p. 35.

it less than a Turdus. Therefore, since I know no other little bird which sits upon the rushes and the reeds, save the Reed Sparrow of the English, I believe that kind to be the Junco. Now this bird is small, a little smaller than a Sparrow, with a longish tail, and a black head. The other parts are brown.

Of the Lingulaca, from Aristotle.

Lingulaca, in Greek called γλωττίς, puts forth a long tongue, whence comes its name; it is one of the leaders of Coturnices, it has the form of a lake-haunting bird.

Of the Lagopus, from Pliny.

The Lagopus is in flavour excellent, its feet shaggy as in a hare have given it this name. Otherwise it is white, in size as the Columbi; it is not eaten except in the land of which it is a native, since it is not tameable while living, and when killed its flesh soon putrefies. There is another bird of the same name, differing but in size from the Coturnices, most excellent for food with yellow saffron sauce.

Of this Martial makes mention in the following verse:—

If my Flaccus delights in the eared lagopes.

Of the Ligurinus or Spinus.

'Ακανθὶς, spinus, or ligurinus, in English a grene finche, as I suppose, in German eyn kirsfincke.

Aristotle.

The Ligurini, commonplace in mode of life and colour, yet excel in pleasantness of song. And they are of the list of birds which feed on thistle-seed. The Florus, the Spinus, and the Ægithus shew mutual dislike. The Spinus wages war moreover with the Ass.

Spinum Ariſtotelis grenefincam noſtram eſſe arbitror: nam illa inter ſpinas plurimùm degit, & ex herbarum ſeminibus uictitat. Auis, quam ſpinum eſſe iudico, magnitudine paſſerem æquat, tota uiridis eſt, præſertim mas in hoc genere, fœmina ferè pallida eſt. ueſcitur carduorum maiorum ſemine & lapparum, ut auriuittis minorum, nidulatur in ramis ſalicũ aut prunorum ſylueſtriũ. cantat amœnè, & cibum & potũ è ſitulis haurire non recuſat.

Sed obijciet mihi forſan quiſpiam, hanc colore uiridi adeoq; amœno, non poſſe ſpinum eſſe, quem Ariſtoteles colore ignobilem eſſe teſtatur. Sciat uelim, qui hoc mihi obijcit, eundem Ariſtotelem uiridem colorem damnare, etiam in aue, quæ tota uiridis eſt, & à uiriditate nomen accepit. Verba Ariſtotelis[1] *hæc ſunt:* Vireo[2], qui totus uiridis eſt, docilis & ad uitę munera [p. 87] ingenioſus notatur, sed malè uolat, nec grati eſt coloris. *Hæc Ariſtoteles.*

DE LVTEA.

Χλώρευς, *luteus ſiue lutea, Anglicè a yelouu ham, a youulryng. Germanicè* eyn geelgorſt.

ARISTOTELES.

Luteus à colore partis ſuæ inferioris pallido dictus, magnitudine alaudæ eſt. Parit oua quatuor aut quinque. Nidum ſibi ex ſymphyto ſtirpitus euulſo facit. Sed ſtragulum ſubijcit ex lana & uillo.

[1] *Hist. An.* Bk IX. 98, 89.

[2] Χλωρίων.

I think that Aristotle's Spinus is our Grenefinc, for it lives for the most part among thorns, and feeds upon the seeds of grasses. The bird which I believe to be the Spinus in its size equals a Sparrow, and is wholly green, and in this kind the male especially, the female being somewhat pale. It feeds upon the seeds of bigger thistles and of burdocks, as the Aurivittis does upon the smaller; and it nests on branches of the willow or wild plum. It is a pleasant songster, and does not refuse to draw its food and water up in little buckets.

But some one may perhaps object to me that this bird with its green and somewhat pretty colour cannot be the Spinus, inasmuch as Aristotle testifies that it is commonplace in colour. I should like the man who thus objects to me to know that Aristotle equally condemns green colour even in a bird which is entirely green, and from its greenness has received its name. These are the words of Aristotle:—

"The Vireo, which is entirely green, is singled out as easy to be taught, and clever for the business of life; but it flies badly and its colour is unpleasing."

So far Aristotle.

Of the Lutea.

Χλωρεύς[1], luteus or lutea, in English a yelow ham, a yowlryng, in German eyn geelgorst.

Aristotle.

The Luteus, so named from the pale colour of its lower parts, is of the size of an Alauda. And it lays four eggs, or even five. It builds itself a nest of comfrey torn up by the roots, but spreads within a covering of wool and hair.

[1] Aristotle in his *History of Animals* mentions three birds, χλωρὶς, χλωρίων, and χλωρεὺς: but Turner's quotation with regard to χλωρεὺς is found in Aristotle under χλωρὶς in a passage (*Hist. An.* Bk IX. 83) where there seems to be no alternative reading.

Auicula, quam luteum esse credo, passere paulò maior est. Maris pectus & uenter lutea sunt: fœminæ uerò pectus luteum, & uenter pallidus est, in capite dorso & alis, pennis fuscis luteæ intermiscentur. Rostrum utrique firmum & breue, in quo tubercum quoddam dentem mentiens, reperias, præter uermes, hordeo & auena libenter uescitur. Cauda huius auiculæ longiuscula est, & frequenter motitans.

DE LVTEOLA.

Χλώρις, *luteola, Anglicè a siskin, Germanicè* eyn [p. 88] zeysich, *quibusdam* eyn engelchen.

Luteola, lutea superiùs descripta, multò minor est, & colore ad uiriditatem magìs tendente, pectore luteo est, & rostro longiusculo, tenui & acuto, auriuittis simili, duas habet maculas nigras: alteram in fronte, alteram sub mento, cantillat non insuauiter. Rara apud Anglos hæc est, nec uspiam ferè alibi quàm in caueis cernitur. Semel tamen in Cantabrigianis agris uidisse recordor. Huius generis sunt, quas Anglia aues canarias uocat.

DE LVSCINIA.

Αἰδών, *luscinia, philomela, Anglicè a nyghtyngall, Germanicè* eyn nachtgäl.

Aristoteles[1].

Parit luscinia æstate quinq; aut sex oua, conditur ab autumno usq; ad uernos dies, luscinia[2] canere solet assiduè diebus ac noctibus quindecim, cùm sylua[3] fronde incipit opacari. dein canit quidem, sed non assiduò, mox adulta ęstate uocem mittit diuersam, nõ insuper ua- [p. 89] riam, aut celerem[4] modulatamq;, sed simplicem,

[1] *Hist. An.* Bk. v. 31.

[2] *Hist. An.* Bk ix. 255.

[3] Aristotle has ὄρος (mountain) here.

[4] Or τραχεῖαν = harsh.

The little bird, which I believe to be the Luteus. is somewhat bigger than a Sparrow. It is yellow on the breast and belly in the cock; but in the hen the breast is yellow and the belly pale. Yellow are mixed with dark feathers upon the head, back, and wings. In each of them the beak is short and stout, and on it one may find a sort of knob that simulates a tooth. Apart from worms it eats barley and oats freely. The tail of this small bird is rather long and is in constant motion.

OF THE LUTEOLA.

Χλωρίς, luteola, in English a siskin, in German eyn zeysich, or of some eyn engelchen.

The Luteola is much smaller than the Lutea above described, and with a colour tending more to green. It has a yellow breast, a longish, slender, pointed bill, like that in Aurivittis, and two spots of black, one on the forehead, one beneath the chin; it warbles with some sweetness. In England it is rare, and scarcely to be seen elsewhere than in cages. Yet I remember having seen it once among the fields of Cambridgeshire. Of this kind are those which England calls Canary birds[1].

OF THE LUSCINIA.

'Αηδών, luscinia, philomela, in English a nyghtyngall, in German eyn nachtgäll.

ARISTOTLE.

In summer the Luscinia lays five or six eggs, but from autumn it lies hid continually until the days of spring. Now the Luscinia is wont to sing incessantly for fifteen days and nights, when woods begin to become dark with foliage. Later it sings indeed, but not incessantly, then in the height of summer it gives forth a different note, not varied over and above, or quick and modulated, but a simple

[1] Gesner, the first to describe the Canary-bird, states that Turner informed him of it.

colore etiam immutatur, & quidem in terra Italia per id tempus alio nomine appellatur, apparet non diu, abdit enim ſeſe & latet.

Ariſtoteles præter unam notam nullam oſtendit peculiarem, qua ab alijs auibus luſcinia differret, ea autem eſt quòd linguæ ſummæ acumine careat. Quanquam & hoc etiam cum atricapilla commune habet. Colore luſcinia, & corporis magnitudine auiculam illam proximè refert, quam Angli lingettam, & Germani paſſerem gramineum nominant. Paſſere paulò minor eſt, & tenuior, & longiori corporis figura, color pectoris ferè cinereus eſt, cætera ſubfuſca.

Graeſmuſch.

DE MERGO.

Αἴθυια, *mergus, Anglicè a cormorant, German.* eyn bůcher.

ARISTOTELES[1].

Mergus marina auis eſt, ex piſcium uenatu uictitans, ſubit tamẽ altiùs in fluuios. Mergus & gauia[2] ſaxis maritimis oua bina aut terna pariunt. Sed gauiæ æſtate, mergi à bruma, ineunte uere. Incubant more cæterarum auium, ſed neutra earum auium conditur.

[p. 90]

Mergus.

Mergus, auis eſt magnitudine ferè anſeris pulla, roſtro longo & in fine adunco, palmipes eſt, & corpore graui, forma corporis aui ſedenti, erecta eſt. Plinius in arboribus nidulari ſcribit, at Ariſtoteles in ſaxis maritimis. Quod uterq; aut uidit, aut à referentibus aucupibus didicit, ſcripto mandauit. Et ego utrumque obſeruaui, nam in rupibus marinis iuxta hoſtiũ Tinæ fluuij mergos nidulantes uidi, & in Northfolcia cum

[1] *Hist. An.* Bk I. 6; Bk VIII. 48, freely rendered.
[2] *Hist. An.* Bk V. 30.

sound. It also changes colour, and during that time is known, at least in the land of Italy, by another name; it is not seen for long, since it conceals itself, and so lies hidden.

Aristotle provides no special mark, save one, by which the Luscinia differs from the rest of birds, and that is that it lacks the point at the tip of the tongue. Though even this it has in common with the Atricapilla. In colour and in size of body the Luscinia comes nearest to that little bird which Englishmen call Lingett and the Germans Grass-Sparrow. It is a little smaller than a Sparrow and more slender, with a longer shape of body, and the colour of the breast is nearly grey; the other parts are brownish.

OF THE MERGUS.

Αἴθυια, mergus, in English a cormorant, in German eyn důcher.

ARISTOTLE.

The Mergus is a sea-bird, and it lives by hunting fishes, yet it makes its way somewhat far up the rivers. The Mergus and the Gavia lay two or three eggs each upon rocks in the sea, the Gaviæ in summer and the Mergi when the spring arrives after the solstice. They incubate like other birds, but neither of these birds conceals itself.

The Mergus, a sad-coloured bird, is nearly equal to a Goose in size, with the bill long and hooked at the end; it is web-footed, heavy in the body, and the attitude is upright in the sitting bird. Pliny writes that it nests on trees, but Aristotle says on sea-rocks. What each man saw or learnt from the reports of bird-catchers he has set down in writing. And I have observed both birds myself, for I have seen Mergi nesting on sea-cliffs about the mouth of the Tyne river, and on lofty trees in Norfolk with the

ardeis in excelſis arboribus. Qui in rupibus maritimis nidificant, ex præda marina ferè uiuũt, qui uerò in arboribus, amnes, lacus, & fluuios, uictus cauſa petunt.

De merope ex Aristotele[1].

Merops. Sunt, qui meropes genitorum ſuorum ſenectutem educare confirmant, uicemq́ꝫ reddi, ut pa- [p. 91] rentes non modò ſeneſcentes, uerùm etiã cùm iam datur facultas, alantur opera liberorum: nec matrem aut patrem exire, ſed in cubili manentes, paſci labore eorum, quos ipſi genuerunt, enutrierunt, educarunt. Pennæ huius auis inferiores pallidę ſunt, ſuperiores cœruleæ ſunt ut halcyonis: poſtremæ pinnulę rubrę habentur. Parit ſex aut ſeptem æſtate in præcipitijs mollioribus, intrà uel ad quatuor cubita ſubiẽs, terræ etiã cauernas ſubiens, cunabula facit.

Plinius[2].

Nec uerò ijs minor ſolertia, quæ cunabula in terra faciunt, corporis grauitate prohibente ſublime petere. Merops uocatur, genitores ſuos reconditos paſcens, pallido intus colore penna- [p. 92] rum, ſuperne cyaneo, priori ſubrutilo. Nidificat in ſpecu, ſex pedum defoſſa altitudine.

Meropem ingenuè fateor me nunquam uidiſſe, nec quẽquam cõueniſſe, qui aliquando uiderit. Tametſi non ſum neſcius apud Germanos, grammaticos non indoctos, eſſe, qui grunſpechtum ſuum, meropem eſſe doceant: ſed Ariſtotele & Plin. reclamantibus. Picus uiridis nidum

[1] *Hist. An.* Bk IX. 82, freely rendered.
[2] *Hist. Nat.* Lib. X. cap. xxxiii.

Herons[1]. Such as make their nests on sea-cliffs generally live on prey from the sea, but such as breed on trees seek rivers, lakes, and streams to get their food.

Of the Merops, from Aristotle.

There are some who insist that Meropes foster the old age of their parents and thus take their turn, so that the parents not in age alone are nourished by the labour of their offspring, but as soon as power is given to these: that neither does the mother-bird fare forth nor yet the father, but they stay within a resting place and are fed by the aid of those which they themselves have bred, nourished and reared. The plumage of this bird is pale beneath, but blue above like that in Halcyon: the pinnules at the end of the wings are reckoned red. It lays six or seven eggs in summer in the softer banks, and makes its nurseries by boring into these for quite four cubits, and it also uses hollows in the soil.

Pliny.

Nor truly is less skill shewn by those birds which make their nurseries in the soil, since the weight of their bodies hinders them from mounting to a height. The kind called Merops feeds its parents in retreat; the colour of its feathers underneath is pale, the upper surface blue, the former being somewhat red. It breeds within a hole, bored out six feet in depth.

In fairness I admit that I have never seen the Merops, nor have I met anyone who ever saw it. Still I am not unaware that there are not unlearned schoolmasters among the Germans, who would teach us that their grunspecht is the Merops, though against the sense of Aristotle and Pliny.

[1] Compare with this Sir T. Browne's *Notes and Letters on the Natural History of Norfolk* (ed. Southwell) p. 11 (1902).

ſibi roſtro ſuo in arboribus facit: ubi enim picus arborem tundēs, illam ex ſono ſubcauam eſſe depræhendit, inſtante tempore partus, eam in qua poſtea nidulaturus eſt, roſtro perforat. Nulla uſpiam arbor tam alta eſt, quam impediente ulla corporis grauitate, non uolatu traijcere poſſit. Pennæ huius quoque ſuperiores ſunt uirides, inferioresq;, niſi malè memini, luteæ aut ſaltem pallidæ ſunt. quare quum merops prohibente corporis grauitate, in ſublime petere, atq; ideo in arboribus nidulari non poſſit, & ſuperne colore ſit cyaneo, Germanorum picus uiridis, quem Britanni à faciendis foraminibus, huholam nominant, merops Ariſtotelis & Plinij eſſe non poterit.

Meropem non eſſe Germanorum grunſpechtum.

DE MERVLA.

Κοττυφός[1], *merula, Anglicè a blak oſel, a blak byrd, Germanicè* eyn merl, *aut* eyn amſel.

[p. 93]

Aristoteles[2].

Merularum duo ſunt genera: alterum nigrum & uulgare: alterum candidum, magnitudine quidem compari, & uoce ſimili, ſed circa Cylenam Arcadię familiare, nec uſquā alibi naſcens. Eſt etiam ex hoc genere, quæ ſimilis nigræ eſt, ſed fuſca colore, & magnitudine paulò minor, uerſari hæc in ſaxis & tectis ſolita eſt, nec roſtrum rutilum, ut merula habet. Merula[3] etiam & colore, & uoce per tempora immutatur. Nam ex nigra redditur rufa, & uocem emittit diuerſam. Strepitat enim per hyemem, quum per æſtatem tumultuans cantet.

Plinius[4].

Merula, ex nigra ruffeſcit, canit æſtate, [p. 94] hyeme balbutit, circa ſolſtitium muta, roſtrum

[1] A misprint for κόττυφος.
[2] *Hist. An.* Bk IX. 95.
[3] *Hist. An.* Bk IX. 254; freely rendered.
[4] *Hist. Nat.* Lib. X. cap. xxix.

Now the Green Picus makes itself a nest with its own bill in trees: for when a Picus hammering on a tree discovers by the sound that it is hollow at the core, the breeding season being close at hand, it bores that with its bill in which it afterwards intends to nest. There is not anywhere a tree so tall which this bird cannot reach by means of flight, for any weight of body that it has. Its plumage is moreover green above and, if my memory serves me, yellow underneath, or pale at least. Since then the Merops, hindered by its weight of body is incapable of rising to a height, and thus of making nests in trees, and has blue upper parts, the grunspecht of the Germans, which the Britons from the holes it makes call huhol [that is, Hew-hole], cannot be the Merops known to Aristotle and Pliny.

Of the Merula.

Κόττυφος, merula, in English a blak osel, a blak byrd, in German eyn merl or eyn amsel.

Aristotle.

Of Merulæ there are two sorts, one black and common, and the other white, of equal size indeed and having a like voice, but which is well-known round Cyllene in Arcadia, and not bred elsewhere. There is of this kind another also, which is like the black, but dull in colour and a little less in size. It usually haunts rocks and roofs, but has not the bill ruddy like the Merula. The Merula in colour and in voice moreover changes with the season, for it turns from black to rufous, and utters a different cry. For it chatters in winter, but sings lustily in summer[1].

Pliny.

From black the Merula turns rufous, in summer it sings, but in winter it babbles, and about the solstice

[1] The readings in Aristotle differ considerably. 'Sings lustily' may go with 'in winter.'

quoq; anniculis in ebur transfiguratur, dũtaxat maribus.

DE MILVO SIVE
miluio.

ἴκτινος, *miluus, Anglicè, a glede, a puttok, a kyte, Germanicè* eyn weye.

PLINIUS[1].

Milui ex accipitrũ genere ſunt, magnitudine differentes. Iidem uidentur artem gubernandi docuiſſe, caude flexibus, in cœlo monſtrante natura, quod opus eſſet in profundo. Milui & ipſi hybernis menſibus latent, non tamen ante hirundines abeuntes. Traduntur & ſolſtitijs affici podagra.

ARISTOTELES[2].

Milui pariunt bina magna ex parte, interdum & terna, totidemq; excludunt pullos. Sed [p. 95] qui Aetolius[3] nuncupatur, uel quaternos aliquando excludit.

Duo miluorum genera noui, maius & minus: maius colore propemodum ruffo eſt, apud Anglos frequens, & inſigniter rapax. Pueris hoc genus cibum è manibus in urbibus & oppidis eripere ſolet. Alterum genus eſt minus, nigrius, & urbes rariùs frequentans. Hoc genus ut in Germania ſæpiſſimè, ita in Anglia nunquam me uidiſſe recordor.

DE MOLLICIPITE.

Μαλακοκρανεύς, *molliceps, Anglicè a shrike, a nyn murder, Germanicè* eyn nuin mürder.

ARISTOTELES[4].

Molliceps eodem in loco ſemper ſibi ſedem ſtatuit, atque ibidem capitur. Grãdi & cartila-

[1] *Hist. Nat.* Lib. X. cap. x.
[2] *Hist. An.* Bk VI. 38.
[3] Other readings are αἰγωλίος and ἐγώλιος.
[4] *Hist. An.* Bk IX. 98.

it is dumb. In yearlings furthermore the bill puts on a look of ivory, provided they are males.

OF THE MILVUS OR MILVIUS.

ἴκτινος, milvus, in English a glede, a puttok, a kyte, in German eyn weye.

PLINY.

Milvi are of the race of Accipitres, though differing in size. They seem, moreover, to have taught mankind the art of steering, by the turning of the tail, nature thus shewing in the sky what might be useful in the sea. Milvi lie hidden in the winter months, yet not until Hirundines depart. They are reported also to be affected with the gout about the solstice.

ARISTOTLE.

Milvi lay for the most part two eggs each, but sometimes three, and hatch as many young. But that kind which is named Ætolian at times lays even four.

I know two sorts of Kites, the greater and the less; the greater is in colour nearly rufous, and in England is abundant and remarkably rapacious. This kind is wont to snatch food out of children's hands, in our cities and towns. The other kind is smaller, blacker, and more rarely haunts cities. This I do not remember to have seen in England, though in Germany most frequently.

OF THE MOLLICEPS.

Μαλακοκρανεύς, molliceps, in English a shrike, a nyn murder, in German eyn nuin můrder.

ARISTOTLE.

The Molliceps invariably takes its stand in the same place, and thereat it is caught. It has a big

gineo capite eſt, magnitudine paulò minor quã turdus, ore firmo, paruo, rotũdo, colore totus cinereo depes[1], & pẽnis inualens eſt, capitur maximè noctua.

Mollicipitem eſſe arbitror auiculam, quam Germani nuinmurder non ſine cauſa nominant. Porrò ut omni- [p. 96] *bus perſpicuum ſit, quæ'nam & qualis illa ſit, formam auis & mores quanto licebit compendio perſtringam. Magnitudine, minimum turdorum genus æquat, è longinquo contẽplanti, tota apparet cinerea. Propius autem inſpicienti, mentum, pectus & uenter alba apparent, ab utroque oculo ad collum uſque, longa & nigra macula, ſed nonnihil obliqua porrigitur. Capite tam grandi eſt, ut aui triplo maiori (modò roſtrum longius & maius eſſet) proportione ſua ſatis reſponderet. Roſtro nigro eſt, & mediocriter breui, & in fine adunco, ſed omnium firmiſſimo & fortiſſimo eſt, utpote quo manum ſemel meam duplici chirotheca munitam, ſauciauerit, & auium oſſa & capita confringat & conterat quàm ocyſſimè. Ala utraque nigra tota eſt, niſi quòd alba linea grandiuſcula, mediam utrinque alam tranſuerſim diſtingat. Caudam picæ ſimilem habet, lõgiuſculam nimirum, & uariam. Tibias & pedes pro ratione corporis omnium minimos, & eos nigros habet. Alas habet breues, & ueluti per ſaltus ſurſum atque deorſum uolitat. Viuit ex ſcarabeis, papilionibus, & grandioribus inſectis: ſed non ſolis iſtis, uerumetiam, more accipitris, auibus. Occidit enim regulos, fringillas, & (quod ego ſemel uidi) turdos. Tradunt etiam aucupes hanc picas quaſdam* [p. 97] *ſylueſtres interdum iugulare, & cornices in fugam adigere. Aues, quas occidit, non unguibus, ut accipitres, uolando perniciter adſequitur, ſed ex inſidijs adoritur, & mox (quod iam ſæpiùs expertus ſum) iugulum petit, & cranium roſtro comprimit & confringit. Oſſa comminuta & contuſa deuorat: & quando eſurit, tantos carnis bolos in gulam ingerit, quantos rictus oris anguſtia poteſt capere. Præter morem etiam reliquarum auium, quando uberior præda contigit, nonnihil in fu-*

Oſſifraga dici poſſit, ſi eius illi magnitudo adeſſet. Nam nec moribu9, nec

[1] ἄπους. Another reading is εὔπους.

and gristly head, and is a little smaller than a Thrush in size; the bill is strong but small, and curved; in colour it is wholly grey, while it is weak-footed and feeble on the wing, it is caught chiefly by the Noctua.

The Molliceps I think to be that little bird which Germans call nuinmurder, not without a cause. Further that it may be quite clear to all which and what sort of bird it really is, I will touch on its form and habits as compendiously as may be. In size it equals the least of the Thrushes, and to one observing from afar seems wholly grey. And yet, to one inspecting it more nearly, the chin, the breast and belly appear white, and from each eye there reaches to the neck, although somewhat oblique, a long black patch. It has so big a head that (were the bill longer and larger) it assuredly would answer in proportion for a bird of thrice its size. The bill is black and moderately short, and hooked at the tip, but is the stoutest and strongest of all, so much so that the bird once wounded my hand, although protected by a double glove, and very speedily it crushes and breaks up the bones and skulls of birds. Each wing is wholly black, except that a white line of some size marks transversely the middle of the wing on either side. The tail is like that of a Pie, that is to say, longish and particoloured. Of all it has the shortest legs and feet proportionately to its body, and these parts are black. It has short wings, and flies as if by bounds upwards and downwards. It lives on beetles, butterflies, and biggish insects, and not only these, but also birds after the manner of a Hawk. For it kills Reguli and Finches and (as once I saw) Thrushes; and bird-catchers even report that it from time to time slays certain woodland Pies, and can put Crows to flight. It does not seize the birds it kills with its claws, after a swift flight, as Hawks do, but attacks them stealthily and soon (as I have often had experience) aims at the throat and with its beak squeezes and breaks the skull. Then it devours the crushed and bruised bones, and when anhungered crams into its gullet lumps of flesh as big as the gape's narrowness can take. Again, beyond the habit of the rest of birds, when prey happens to be more plentiful, it lays by some for future scarcity.

colore ab ea multùm abludit.

turam penuriã reponit. Muſcas enim grandiores & inſecta iam capta in aculeis & ſpinis arbuſtorum figit & ſuſpendit: omnium auium facilimè cicuratur, & manſuefacta, carnibus alitur, quæ ſi fuerint ſicciores, aut prorſus exangues, potum requirit. In Anglia ſæpiùs quàm bis nunquam uidi, in Germania ſæpiſſimè. Nomen huius apud noſtros neminem inueni, qui nouerit, præter Dominum Franciſcum Louellum, tam animi quàm corporis dotibus equitem auratum nobiliſſimum. Iam ſi cui mollicipitis Ariſtotelis deſcriptio huic non uideatur per omnia conuenire, tyrannorum albo adſcribat, aut auem oſtendat, cui deſcriptio meliùs competat.

DE NOCTVA.

Γλαύξ, *noctua, Anglicè an ouul, or an houulet,* [p. 98] *Germanicè* eyn eul & eyn ůle *Saxonicè.*

ARISTOTELES[1].

Noctuæ, cicuniæ[2], & reliqua, quę interdiu nequeunt cernere, noctu uenando cibum ſibi adquirunt: uerùm non tota nocte id faciunt, ſed tempore ueſpertino et matutino. Venantur autem mures, lacertas, uerticillos, & eiuſmodi beſtiolas. Noctuam[3] cæteræ omnes aues circumuolant, quod mirari uocatur, aduolantesq; percutiunt[4]. Qua propter aucupes ea conſtituta, auicularum genera multa & uaria capiunt.

DE OLORE.

Κυκνὸς, *olor, Anglicè a ſuuan, Germanicè* eyn ſwān.

ARISTOTELES[5].

Olor.

Olores palmides[6] ſunt, apud lacus & paludes uiuentes, qui nec probitate uictus, morum, pro- [p. 99] lis, ſenectutis uacant[7]. Aquilam ſi pugnam

1 *Hist. An.* Bk IX. 122.
2 Apparently a misprint for cicumæ, said to mean 'horned owls.' Aristotle has νυκτικόρακες which he identifies in Bk VIII. 84 with ὦτοί.
3 *Hist. An.* Bk IX. 11.
4 Aristotle has τίλλουσι = pluck it.
5 *Hist. An.* Bk IX. 78.
6 That is, palmipedes.
7 Aristotle has εὐβίοτοι δὲ καὶ εὐήθεις καὶ εὔτεκνοι καὶ εὔγηροι.

For it impales and hangs the bigger flies and insects on the thorns and spines of shrubs, so soon as they are caught: of all birds it is tamed most easily, and when accustomed to the hand is fed on meat, and, should this happen to be somewhat dry or altogether bloodless, it requires drink. In England I have never seen it oftener than twice, although most frequently in Germany. Among our people I have found no one who knew its name, except Sir Francis Lovell, that most noble knight, endowed with equal gifts of mind and body. Now if Aristotle's description of the Molliceps does not appear to any one in all points to agree with this let him ascribe it to the list of the Tyranni, or shew us a bird, which the description fits better than this.

OF THE NOCTUA.

Γλαύξ, noctua, in English an owl or an howlet, in German eyn eul, and in Saxon eyn ůle.

ARISTOTLE.

The Noctuæ, Cicumæ and the rest, which cannot see by day, obtain their food by seeking it at night: and yet they do not do this all night long, only at eventide and dawn. They hunt moreover mice, lizards, and scorpions, and small beasts of the like kind. All other birds flock round the Noctua, or, as men say "admire," and flying at it buffet it. Wherefore this being its nature[1], fowlers catch with it many and different kinds of little birds.

OF THE OLOR.

Κύκνος, olor, in English a swan, in German eyn swän.

ARISTOTLE.

Olores are web-footed, and they live on lakes and marshes; they get food with ease, are peaceable, prolific and attain to a great age. They repulse the

[1] Or, possibly, 'the bird being set down on the ground.'

cœperit, repugnantes uincunt. Ipſi tamen nunquam, niſi prouocati, pugnam inferunt. Canere ſoliti ſunt, & iamiam morituri. Volant etiam in pelagus longius, & iam quidam cùm in mari Africo nauigarunt, multos canentes uoce flebili & mori nonnullos conſpexêre.

Si quis olorem nunquam uiderit, & ex hac Ariſtotelis deſcriptione non ſatis qualis ſit auis didicerit, ſciat auem eſſe albam, anſere multò maiorem, forma tamen & uictu ſimilem, pedibus nigris, & roſtro parùm turbinato, colore rutilo, in cuius ſumma parte, qua capiti committitur, nigerrimum tuberculum, atque id rotundum, & in roſtrum ſeſe inflectens, exiſtit.

DE ONOCRATALO.

Onocrotalus. *Sunt hodie non parùm multi eruditione inter omnes conſpicui, qui grandiſonam illam lacuſtrem auem, Anglis buttoram & Germanis pittourum, & roſdommam uocatam, Onocrotalum eſſe contendant. Quorũ* [p. 100] *ego ſententiæ lubens ſubſcriberem, (pulchrè enim cum uoce auis nominis etymologia conuenit:) niſi Plinij autoritas de onocrotalo ad hunc modum ſcribentis, non diſſuaderet.* Onocrotali, *inquit,* olorum ſimilitudinem habent, nec diſtare uidentur omnino, niſi faucibus ipſis ineſſet, alterius uteri genus. huc omnia inexplebile animal congerit, mira ut ſit capacitas, mox perfecta rapina, ſenſim inde in os reddita, in ueram aluum ruminantis modo refert. Gallia hos ſeptentrionalis, proximè[1] oceano mittit. *Hæc Plinius*[2].

[1] This should apparently be 'proxima,' as some texts have it.
[2] *Hist. Nat.* Lib. x. cap. xlvii.

Aquila successfully, should he begin a fight; and yet, unless provoked, never induce the fight. These birds are wont to sing even when just about to die. They also fly afar over the main, and men ere now, who have been sailing on the African sea have met with many singing mournfully and seen some of them die.

Should any one have never seen a Swan, nor learnt sufficiently what sort of bird it be from this account of Aristotle, let him know that it is a white bird, much bigger than a Goose, though like in form and feeding; with black feet, and a bill hardly spindle-shaped[1], reddish in colour; on the highest part of which, where it adjoins the head, stands forth a very black and rounded knob, sloping towards the bill.

Of the Onocrotalus.

There are many to-day conspicuous among all for learning to no small degree who maintain that the loud-sounding lacustrine bird, called Buttor by the English, and Pittour or Rosdomm by the Germans, is the Onocrotalus. To whose opinion I would willingly subscribe, (the more so as the etymology of the bird's name agrees well with its voice,) did not the authority of Pliny writing of the Onocrotalus after this manner dissuade me therefrom.

The Onocrotali, he says, have a similitude to the Olores, and they do not seem to differ in any way, save that there is a kind of second belly in the very jaws. Herein the insatiable animal crams everything at once, so marvellous is its capacity, and presently, the plundering complete, it gradually returns all to the mouth, and thence transfers it to the real belly in the manner of a ruminant. Northern Gaul, where nearest to the ocean, sends us these. So far Pliny.

[1] This passage is not easily rendered, as it is difficult to see what Turner intended by 'turbinato.' Turbo is a conical shell, spindle and so forth; but it is hard to say how a Swan's beak could be considered either conical or spindle-shaped.

Nunc paucis auem illam uobis depingam, quã onocrotalum esse asseuerant. Auis est tota corporis figura Ardeæ similis. longis cruribus, sed ardeæ breuioribus. longo collo, & mirè plumoso, & rostro nec breui nec obtuso. caput pennæ tegunt nigerrimæ. reliquum uerò corpus, fuscæ & pallidæ maculis nigris densissimè respersæ. Pedes habet longissimos, nam inter extremos ungues medij digiti pedis unius & calcis eiusdem, [p. 101] *spithames longitudo intercedit. Vngues habet longissimos, nam ille, qui calcis uicem in auibus gerit, longitudine sesquiunciam superat. quare ad fricandos dentes nostrates utuntur, & argento inserunt. Medius digitus utriusque pedis, qui cæteris longior est, unguem habet portentosum, nempe dentatum & serratum, non secus atque pectunculorum testæ serratæ sunt, ad lubricas anguillas, quas cœpit*[1]*, retinendas, à natura proculdubio ordinatum. Cauda illi breuissima est, et stomachus capacissimus, quo ingluuiei loco utitur. Ventriculum non cæterarum auium uentriculis, sed canino similem habet, & eum grandem & capacem.*

Sed ne cui falsa esse uideantur, quæ de hac aue iam scripsi, aut ex aliorum relatu potiùs quàm certa experientia didicisse uidear: dum prima huius libri folia adhuc sub prælo essent, auem mihi hanc contemplanti, secantiq;, & nũ tales haberet uentriculum & stomachum, quales Plinius illi tribuit, inuestiganti: aderant uir eruditissimus, & abstrusiorum naturæ arcanorum studiosissimus inuestigator Ioannes Echthius, Medicus apud Colonienses celeberrimus: Cornelius Sittardus, Medicinæ prima laurea decoratus. M. Lubertus Estius, artium liberalium professor, ambo simplicium medicamentorum pulchrè gnari, & ad miraculum usque studiosi: [p. 102] *& Conradus Embecanus uir non uulgariter doctus, et Gymnicanæ officinæ castigator insigniter diligens, cum alijs aliquot bonarum artium studiosis, qui me nihil de hac aue hîc scripsisse testari possunt & uolunt, quod cum illis omnibus non uiderim. Ad ripas lacuum & paludium desidet, ubi rostrum in aquas in-*

[1] A misprint for 'cepit.'

Now in a few words I will portray to you that bird which they assert to be the Onocrotalus. In general make of body it is like the Heron, with long legs, though shorter than that bird's. The neck is long and marvellously thick with plumes, the beak is neither short nor blunt. Very black feathers clothe the head, but on the body generally they are dusky and pale, and most thickly sprinkled with black spots. It has very long feet, indeed there is a span's length from the claw-tips of the middle toe of either foot to the heel of the same. It has very long claws, for that which serves in birds the purpose of a heel exceeds an inch and a half in length, on which account our countrymen use it to pick their teeth, and mount it in silver. The middle toe of either foot, which is longer than the rest, has a prodigious claw, that is to say, toothed and serrated, not unlike the shells of little scallops are, doubtless contrived by nature to retain the slippery eels, which the bird catches. The tail is very short, the gullet most capacious, and it uses it in the place of a crop. It has a belly not like that of other birds, but like that of a dog; it also is large and capacious. But lest what I have written thus far of this bird seem false to anyone, or lest I seem to have learnt the above from the reports of others rather than from sure experience: while the first pages of this book were still at press, and while I was examining the bird and was dissecting it, and taking note whether it really had a belly and a stomach such as Pliny had assigned to it, there were assisting me Joannes Echthius, a very learned man and a most zealous student of the more abstruse secrets of nature, a physician much renowned among the men of Cullen: Cornelius Sittardus decorated with the highest laurel-wreath of Medicine: Marcus Lubertus Estius, professor of the liberal arts, both excellently skilled in that of simpling, and wonderfully earnest, and as well as these Conradus Embecanus, a man well-informed in no common degree, and a remarkably careful corrector in the printing-house of Gymnicus, with certain others versed in learned arts, who can and will bear witness to the fact that I have written nothing here about this bird which I have not observed in company with all of them. It sits about the sides of lakes and marshes, where putting

ſerens, tantos êdit bombos, ut ad miliarium Italicum facilè poſſit audiri. Piſces & præſertim anguillas uorat auidiſſimè, nec ulla auis eſt, excepto mergo, quæ iſta uoracior eſt. Nunc quid ſimile habet iſta cygno? Nihil planè, quod ſe oculis conſpiciendum offerat. Et Moiſes Leuit. undecimo capite, proximè cygnum inter immundas aues onocrotalum recenſet. Vnde non immeritò ſuſpicio quibuſdam ortà eſt in Gallia, aut Iudæa auem forma olori ſimilem alicubi poſſe reperiri. Quòd ſi nuſquam talis inueniatur: probabile eſt, aut Plin. à mendacibus relatoribus ſuis eſſe falſum, aut ea, quæ de ſimilitudine inter onocrotalum & cygnũ tradidit, non de corporis ſed uocis ſimilitudine intellexiſſe. Nam & olores interdum bombos emittunt ruditui aſinino non diſſimiles: ſed breues, & quæ longè audiri non poſſunt. Verùm ſi hanc meam interpretationẽ uariæ, reconditæq́; eruditionis uiri, ſuis ſuffragijs minimè approbauerint, [p. 103] *hanc ſaltem Ariſtotelis ardeam ſtellarem eſſe mecum conſentient. Nam præter cætera, quæ ſuperiùs attigi, Ariſtoteles in fabula fuiſſe oſtendens, ardeam ſtellarem ex ſeruo auem fuiſſe factam, opinioni meæ multum patrocinatur. Vt fugitiuorum enim ſeruorũ poſt fugam depræhẽſorum, cutis, loris, flagris, uirgis, & ſcorpionibus icta, uerberum uibicibus, tota maculoſa redditur: ita huius auis plumæ nigris ubique maculis, ſed potiſſimùm in tergo, diſtinctæ & ueluti picturatæ, ſerui flagris cæſi cutem proximè referũt. Quam rem fabulæ occaſionem dediſſe ex hoc colligo, quòd fabularum uariarum autor Ariſtophanes*[1], *de attagene aue, quod ad plumarum colores attinet, huic ſimilima, ad hunc modum ſcribat:*

Onocrotalum quibuſdam hodie dictũ, ardeam, eſſe ſtellarem apud Ariſtotelem.

Si quis ex uobis erit fugitiuus atq; uſtis notis,
Attagen is ſanè apud nos uarius appellabitur.

De ortygometra ex Aristotele[2].

Ortygometra, id eſt, coturnicũ matrix, auis eſt forma perinde ac lacuſtres. Cruribus ideo

[1] See p. 36.
[2] *Hist. An.* Bk VIII. 83, freely rendered, and interpolated.

its beak into the water it gives utterance to such a booming as may easily be heard an Italian mile away. It gorges fishes and especially eels most greedily, nor is there any bird, except the Mergus, that devours more. Now what resemblance has it to a Swan? Distinctly none that brings itself in view before our eyes. Now in the eleventh chapter of Leviticus Moses enumerates the Onocrotalus next to the Swan among the unclean birds. And a suspicion has arisen thence, not undeservedly, within a certain class, that somewhere within Gaul or Judæa a bird of Swan-like form may possibly exist. If such, however, nowhere can be found, it seems likely that Pliny either was deceived by lying story-tellers or he understood that which he has related of the similarity between the Onocrotalus and Cygnus to refer to a resemblance not of body, but of voice. For even Swans utter at certain times booms not unlike the braying of an ass: but short, and which cannot be heard afar. However if men of deep and varied learning by their votes shall not approve this rendering of mine, at least they will agree with me that the said bird is Aristotle's Ardea Stellaris. For to omit the rest, which I have touched upon above, that author certainly gives countenance to my opinion when he shews a tale to have existed that the Ardea Stellaris from a slave was turned into a bird. For as the skin of an absconding slave, caught subsequent to flight, stricken with thongs, whips, rods and knotted ropes, becomes all mottled with the wales of stripes, so too the feathers of this bird are marked, and painted as it were, with mottlings of black in every part, though chiefly on the back, and thus may well recall to us the skin of slaves cut up with whips. And that this thing gave rise to the aforesaid tale, I gather from the fact that Aristophanes, author of various plays, writes of the Attagen, a bird very like ours so far as colour of the feathers goes, to this effect:—

"If any of you be a runaway, and branded with the marks, with us assuredly he shall be called the spotted Attagen."

Of the Ortygometra from Aristotle.

The Ortygometra, that is, dam of the Coturnices, in form is much like marsh-birds. Certain birds are

longis aues quædam innituntur, quòd earum uita ſit paluſtris.

[p. 104] *Ortygometram aliqui eandem eſſe auem cum crece et cychramo uolunt. Sed Ariſtoteles, peculiare caput creci donauit, et octauo libro hiſtoriæ animalium, cychramum a matrice, quam ortygometram uocat, his uerbis diſtinguit.* Coturnices (*inquit*) cùm hæc adeunt loca, ſine ducibus pergunt: at cùm hinc abeunt, ducibus lingulaca, oto, & matrice, proficiſcuntur, atque etiam cychramo, à quo etiam reuocãtur noctu, cuius uocẽ cùm ſenſerint aucupes, intelligunt parari diſceſſum. *Hæc ille.*

Fieri igitur non poteſt, ut matrix & cychramus eadem auis ſit. Aliqui ortygometram eſſe uolunt Germanorum ſcricam, & Anglorum daker hennam, quorum ego ſententiæ accederem, ſi crecem eandem cum iſta, euincerent.

DE OSSIFRAGA.

Aristoteles[1].

Oſſifragę magnitudo maior eſt quàm aquilæ, color ex cinere[2] albicans. Probè[3] & fœtificat, & [p. 105] uiuit, cœnæ gerula & benigna eſt. Nutricat enim bene, & ſuos pullos & aquilæ. Cùm enim illa ſuos nido eiecerit, hæc recipit eos, ac educat.

Plinius[4].

Quidam adijciunt genus aquilę, quam barbatam uocãt Thuſci oſſifragam.

[1] *Hist. An.* Bk VIII. 39.
[2] Apparently a misprint for 'cinereo.'
[3] *Hist. An.* Bk IX. 123.
[4] *Hist. Nat.* Lib. X. cap. iii.

perched upon long legs because their life is passed in marshes.

Some will have Ortygometra to be the same as Crex and Cychramus. But Aristotle has attributed a peculiar sort of head to Crex, and in the eighth book of his History of Animals distinguishes his Cychramus from Matrix, which he calls Ortygometra, in the following words:—

Coturnices (he says) when they come to these places travel without guides: but when they go away set out with the Lingulaca, the Otus, and the Matrix as their guides, and also with the Cychramus, by which they are moreover summoned back at night. And when the fowlers have heard its cry, they know the birds' departure is at hand.

Thus he writes.

Therefore it is impossible that the Matrix and the Cychramus should be the same. Others will have the Ortygometra to be the Scrica of the Germans and the Daker Hen of the English, and I should accede to their opinion, if they could but prove the Crex to be the same as this.

OF THE OSSIFRAGA.

ARISTOTLE

In size the Ossifrage is greater than the Aquila, its colour whitish grey[1]. Both in breeding it is comely and in way of life, it brings food home and is kindly. For it rears its own young with care, besides those of the Aquila. For when the latter has cast its progeny out of the nest, the former takes them to itself, and brings them up.

PLINY.

Some there are who add that kind of Aquila, which the Tuscans call a bearded Ossifrage[2].

[1] See p. 36.

[2] Possibly Pliny means the Lämmergeier (*Gypaëtus barbatus*).

DE OTO.

ὠτός, otus, Anglicè a horn oul, Germanicè eyn ranſeul / oder eyn ſchleier eul.

ARISTOTELES [1].

Otus noctuæ ſimilis eſt, pinnulis circiter aures eminentibus, præditus, unde nomen accepit, quaſi auritum dixeris. Nonnulli ululam eũ appellant, alij aſionem [2]. Blatero hic eſt & hallucinator, & planipes: ſaltantes enim imitatur. Capitur intentus in altero aucupes [3], altero circumeunte.

[p. 106]

DE OTIDE EX PLINIO [4].

Tetraonibus proximæ ſunt, quas Hiſpania aues tardas appellat, Græcia otidas, damnatas in cibis. Emiſſa enim oſſibus medulla, odoris tædium extemplò ſequitur.

DE PARIS.

αἰγιθαλός, parus, Anglicè a tit mouſe, German. eyn meyſe.

ARISTOTELES [5].

Parorum tria ſunt genera: fringillago, quæ maior eſt, quippe quæ fringillam æquet. Alter monticola cognomine eſt: quoniam in montibus degat, cui cauda longior. Tertius magnitudine ſui exigui corporis diſcrepat, quanquam cætera ſimilis eſt. parus [6] plura oua parit.

Fringillago.

Primum parũ, Angli uocant the great titmous or the great oxei, Germani eyn kölmeyſe.

Parus medius.

Parum ſecundum, Angli the leſs titmous nominant. Germanici eyn meelmeyſe.

1 *Hist. An.* Bk VIII. 84. very freely rendered.

2 Aristotle has 'νυκτικόρακα,' instead of 'ululam' and 'asionem.'

3 A misprint for 'aucupe.'

4 *Hist. Nat.* Lib. X. cap. xxii.

5 *Hist. An.* Bk VIII. 40.

6 *Hist. An.* Bk IX. 88.

Of the Otus.

ὠτός, otus, in English a horn owl, in German eyn ranseul or eyn schleier eul.

Aristotle.

The Otus is like a Noctua, furnished with little tufts sticking out near the ears, whence it has got its name, as though one should say "eared." Some call it Ulula, and others Asio. It is a babbler and a mischievous rogue, and is a mimic too, for when men dance it imitates their ways. It is caught while intent upon one of two bird-catchers, the other circumventing it.

Of the Otis from Pliny.

Next to the Tetraones come those birds, which Spain calls "Aves tardæ" and Greece "Otides," condemned as food for man. For when the marrow issues from the bones, disgust at the smell follows there and then.

Of the Pari.

αἰγίθαλος, parus, in English a titmouse, in German eyn meyse.

Aristotle.

There are three kinds of Pari: Fringillago bigger than the rest, for it is equal to a Fringilla. The next Monticola by name, for it inhabits mountains, has a longer tail. The third kind differs in the size of its small body, though not otherwise unlike the rest. Parus lays many eggs,

The first Parus the English call the Great Titmouse or Great Oxeye, the Germans eyn kölmeyse.

The second Parus the English name the Less Titmouse, the Germans eyn meelmeyse.

[p. 107] Parus minimus.

Parum tertium, Angli nonnam à ſimilitudine quam cum uelata monacha habet, nominant.

Nidulantur pari in cauis arboribus, ueſcuntur non ſolùm uermibus, ſed & canabino ſemine, & nucibus, quas roſtris ſuis acutioribus ſolent perforare, & nucleos eruere. Sæuo duo priora genera multùm delectantur. Parus maximus ineunte ſtatim uere cantiunculam quandam breuem, nec admodum iucundam exercet, aliâs mutus, huic pectus luteum eſt, intercurſante linea nigra maiuſcula. Cæterorum corpora albo, nigro, pallido, & cyaneo coloribus diſtinguuntur.

DE PARDALO.

Pardalus, Angl. (ut creditur) a pluuer, Germa. eyn puluer.

Aristoteles[1].

Pardalus etiam auicula quædam perhibetur, quæ magna ex parte gregatim uolat, nec ſingularē hanc uideris, colore tota cinereo eſt, magnitudine proxima mollicipiti[2] eſt: ſed pennis & pedibus bonis, uocem frequentem nec grauem emittit.

[p. 108] *Si auis illa pardalus ſit quam eſſe ſuſpicor, celerrimè currit, & ſibilum, quem paſtores & aurigarum pueri labijs porrectis êdunt, uoce imitatur. Pennas habet ad cinereum colorem proximè uergentes, quarum ſingulæ ſingulis flauis maculis ſunt reſperſæ, & ea auicula, quam mollicipitē eſſe conijcio, multò maior eſt. Fieri poteſt, ut eius auis plures ſint ſpecies.*

DE PASSERIBVS.

στρουθος, paſſer, Angli. a ſparrouu, German. eyn müſche oder eyn ſpätz. Quidam eyn lüningk, *Saxones* eyn ſperlingk *uocant.*

Paſſer, authore Ariſtotele[3], puluerat & lauat, et auis eſt omnium ſalaciſſima. Et quanquam Ariſtoteles unum tantùm paſſerum genus fecerit, tria tamē genera eſſe conſtat, quæ nunquam nouit, & quæ recentiores

[1] *Hist. An.* Bk IX. 99.

[2] Aristotle has ἐκείνοις, referring to the χλωρίων and the μαλακοκρανεύς.

[3] *Hist. An.* Bk IX. 260, Bk V. 8.

The third Parus the English name the Nun from the resemblance that it bears to a veiled sister.

The Pari nest in hollow trees, they feed not only on worms, but on hempseed and nuts, which they are wont to bore with their sharp-pointed beaks, and thence extract the kernels. The two former kinds are very fond of suet. The Greatest Parus, when the spring arrives, at once utters a sort of little song, short and not very pleasing, it is dumb at other times; its breast is yellow with a somewhat big black line running along the middle. Of the other kinds the bodies are diversified by white, black, grey, and blue.

Of the Pardalus.

Pardalus, in English (as is believed) a pluver, in German eyn pulver.

Aristotle.

The Pardalus again is held to be a certain little bird, which for the most part flies about in flocks, and cannot be seen solitary; it is wholly grey in colour, and in size comes nearest to the Molliceps: but it has strong wings and feet, and utters a frequent but not deep-toned cry.

If that bird be the Pardalus which I suspect, it runs very swiftly, and by its cry mimics the whistle which shepherds and post-boys make with pouting lips. It has the feathers almost ash-colour, each sprinkled with one yellow spot, and is much bigger than the little bird which I suppose to be the Molliceps. It well may be that there are several kinds of this bird.

Of the Passeres.

στρουθός, passer, in English a sparrow, in German eyn müsche or eyn spätz. Some call it eyn lüningk, the Saxons eyn sperlingk.

The Passer, Aristotle says, both dusts itself and washes, and is of all birds most wanton. And though Aristotle has made only one kind of Passeres, yet it is clear that there are three kinds, which he never knew; but which the later

Paſſer torquatus.

Græci inuenerunt. Primum horum trium eſt paſſer torquatus, à communi paſſere, nõ ſolũ torque albo, ſed & uoce, & modo nidificandi differẽs. Hoc genus in Germania frequens eſt, ſed apud Anglos rarum. Secundus paſſer magnus Auctuario dicitur, & in ſummis arborum ramis plerumque ſolet ſedere. hunc uarijs de cauſis Anglorum buntingam, & Germanorum Gerſthammeram eſſe ſuſpicor. Tertius paſſer Ariſtoteli incognitus, eſt paſſer troglodites, apud Paulum Aeginetam, & Aetium celeberrimos medicos, multùm celebratus. Qualis autem illa auis ſit, ex Paulo & Aetio, quorum deſcriptiones mox ſubijciam, omnibus facilè patebit.

Paſſer magnus. Huic mollicipitis apud [p. 109]

Ariſtotelem deſcriptio magna ex parte conuenit.

Paſſer troglodites.

PAULUS AEGINETA DE PASSERE TROGLODITE.

His accedit laudatiſſimum remedium troglodites. eſt autem paſſerculus omnium auium minima, ea ſola excepta, quę regulus appellatur, hoc enim ſolo paulò maior eſt, eiq; ſimilis : colore inter cineriũ & uiride, tenui roſtro, in muris maximè & in ſepibus degens.

AETIUS.

Troglodites eſt paſſerculus minimus, iuxta ſepes & muros uictum quæritans. Eſtq; hoc animalculum omnium auicularum minimum, excepta ea, quæ regulus appellatur, ſimilis autem regulo in multis, præterquam quòd in fronte auricolores pennas non habet. Eſt autẽ troglodites paſſer regulo paulò maior & nigrior, caudamq; ſemper ſubrectam, & albo colore retrò interpunctam habet. Magis item garrulus

[p. 110]

Greeks discovered. First of these three is Passer torquatus, differing from the common Passer not alone in its white collar, but also its note and mode of nesting. This kind is plentiful in Germany, but rare among the English. The second Passer is called in the Supplement[1] the Great, and for the most part it is wont to sit on the top boughs of trees. For several reasons I consider this to be the Bunting of the English and the Gersthammer of Germans. The third Passer, unknown to Aristotle, is the Passer troglodytes, fully recognised by Paulus Ægineta and Aëtius, doctors of great renown. And so what sort of bird it is will easily be seen by everyone from Paulus and Aëtius, and their descriptions I will forthwith add.

PAULUS ÆGINETA[2] OF THE PASSER TROGLODYTES.

There is a remedy most highly prized besides these, namely Troglodytes: this is nothing but a little Sparrow, the least of all birds, with the exception only of that kind which is called Regulus. It is a little bigger than that bird alone, and similar to it: in colour between grey and green, and with a slender bill. It lives chiefly in walls and hedges.

AËTIUS[3].

The Troglodytes is the very least of Sparrows, seeking for its food near hedges and near walls. This little animal moreover is the smallest of all little birds, except that which is called Regulus, while it is like the Regulus in many ways, save that it has not golden-coloured feathers on the forehead. The Passer troglodytes is a little larger and blacker than the Regulus; it always has its tail cocked up, which is spotted behind with white. Likewise it is more noisy than the

[1] It seems impossible to ascertain what this Auctuarium was.

[2] A medical writer of Ægina, whose chief work was *De Re Medica Libri Septem.*

[3] A Greek medical writer of Amida in Mesopotamia who wrote Βιβλία Ἰατρικὰ Ἑκκαίδεκα.

quàm regulus eſt, & ſanè iuxta ſummum alæ lineamentum cinerij amplius coloris. Breues item facit uolatus, naturalem autem uim omnino admiratione dignam habet.

Nihil eſt in hac deſcriptione, quod non ad amuſſim auiculæ conueniat, quam Angli paſſerem ſepiarium, Colonienſes aucupes koelmuſſhum nominant. Sed quoniam tam in Germania quàm in Anglia uarijs nominibus appellatur, & non omnes eum ex uno nomine agnoſcunt, omnia eius, quæ noui nomina, ut omnibus innoteſcat, ſubijciam.

Vocatur apud Anglos an hedge ſparrouu, hoc eſt paſſer ſepiarius, & a dike ſmouler, hoc eſt, in ſepibus [p. 111] *deliteſcens. Vulgus Coloniēſe hunc paſſerem eyn graſsmuſch appellat. uerū peritiores quiq; aucupes eyn koelmuſch, hoc eſt, paſſerem in foraminibus & cauernis degentem, nuncupant. Hîc Germanos monitos uolo, quū duæ ſint aues, graſmuſchi, ſua lingua uocatæ, illā ſolā eſſe trogloditen, quæ per totum annum regulo ſimilis cernitur, & non illam, quæ circa fauces plumoſa*[1] *ineunte ſtatim hyeme diſcedit. Nidum huius paſſeris ſemel humi factum inter urticas uidi, & pullos antequam uolare poſſunt, relicto nido, inter herbas fruticesq; reptitantes, ſæpiùs obſeruaui: uermibus paſcitur, & paulò ante ueſperum ſolet impenſiùs ſtrepere, & omnium ferè auium poſtrema dormitum petit.*

DE PAVONE.

Ταών, *pauo, Anglicè a pecok, Germanicè* eyn pffaw. *Saxonicè* eyn pagelün.

PLINIUS[2].

Pauo gemmantes laudatus expandit colores, aduerſo maximè ſole, quia ſic fulgentiùs radiant, ſimul umbræ quoſdam repercuſſus cæteris, quæ [p. 112] in opaco clariùs micāt, conchata quærit cauda, omnesq; in aceruum contrahit pēnarum, quos

[1] This is probably a misprint for some other word.
[2] *Hist. Nat.* Lib. X. cap. xx.

Regulus, and certainly towards the upper border of the wing the colour is more nearly grey. And though it takes short flights, its natural energy is worthy of all admiration.

In this description there is nothing that does not agree exactly with the little bird, which Englishmen name Passer sepiarius, the bird-catchers of Cullen the koelmusch. But inasmuch as both in Germany and England it is called by various names, and all men do not recognise it by the same, I will subjoin those of its appellations which I know, that so it may be known to all.

By the English it is called a Hedge-Sparrow, which is the same as Passer sepiarius, and also a Dike Smouler, one, that is, hiding itself in hedges. The common people of Cullen call it eyn grassmusch, but those who know better and fowlers name it eyn koelmusch, that is, a Sparrow dwelling within holes and caverns. Now here I wish the Germans to be warned, that since there are two birds called grasmusch in their tongue, the Troglodytes is that kind alone which throughout the year is noticed to be like the Regulus and not that which is feathered round the jaws, and goes away so soon as winter comes. Of this Passer I once met with a nest built on the ground among nettles and I have often seen young having left the nest before that they could fly, creeping among the grass and shrubs. It feeds on worms, and it is wont a little before evening to cry out with not a little vehemence; it goes to roost almost the last of all the birds.

Of the Pavo.

Ταών, pavo, in English a pecok, in German eyn pffaw, in Saxon eyn pagelün.

Pliny.

The Peacock is admired for setting forth his jewelled colours, generally counter to the sun, since thus they shine the brighter, while with concave tail he gains certain reflexions of the shade for other feathers which shine brighter in the dark, and at the same

ſpectare[1] gaudet oculos. Idem cauda annuis uicibus amiſſa çum folijs arborum, donec renaſcatur iterum cum flore: pudibundus ac mœrens quærit latebram. Viuit annis uiginti quinque. Colores incipit fundere in trimatu. Ab autoribus traditur non tantùm glorioſum animal, ſed maleuolum, ſicut anſer uerecundum. Pauones in capitibus ſuis ueluti crinitã habent arbuſculã.

DE PERDICE.

Πέρδιξ, *perdix*, *Anglicè a pertrige*, *Germanicè* eyn velt hön/ ader eyn raphön.

Aristoteles[2].

Perdix auis eſt pulueratrix, & non altiuola, & eadem non in nido ſed in condenſo frutice [p. 113] aut ſegete prolem ſuam munit. Aues enim grauiores nidos ſibi non faciunt, ut coturnices & perdices, & reliquæ generis eiuſdem. Quibus enim uolandi facultas deeſt, ijs nidus non prodeſt: ſed facta in aprico[3], area, (alibi enim nuſquam pariunt) atque materia ut uepribus[4] quibuſdam congeſtis, quoad accipitrum & aquilarum iniuriam deuitare poſſint. Oua edunt, & incubant, mox cùm excluſerint, protinus pullos educunt. propterea quòd nequeunt ſuo uolatu ijs cibum adminiſtrare. Refouent pullos ſuos ſub ſe, ipſæ ducendo more gallinarum, & coturnices & perdices. Nec eodem loco pariũt & incubãt, ne quis locum percipiat, longioris temporis mora.

[1] A misprint for 'spectari.'

[2] *Hist. An.* Bk IX. 260, 59—61.

[3] There is a reading λείῳ besides ἡλίῳ.

[4] Aristotle has *ἄκανθάν τινα καὶ ὕλην*, so no doubt 'ut' is a misprint for 'et.' Gaza has 'ut.'

time draws into a cluster all the eyes upon his feathers, which he is well pleased should be admired. The same bird, having lost his tail, when the trees shed their leaves by annual change, ashamed and sorrowing seeks a hiding place, until it once more grows together with the flowers. He lives for five and twenty years, and in the third begins to shew his colours. He is reported by authorities to be an animal not only proud but also ill-disposed, just as the Goose is bashful. Peacocks have on their heads as it were a bush of hair.

OF THE PERDIX.

Πέρδιξ, perdix, in English a pertrige, in German eyn velt hön, or eyn raphön.

ARISTOTLE.

The Perdix is a bird that dusts itself, and flies not high[1]; moreover it finds safety for its young not in a nest, but in thick shrubs and corn. For birds of heavy body make no nest, such as Coturnices and Perdices, and others of like sort. For to those kinds in which facility of flight is wanting, there is small advantage in a nest; but in some sunny place (for they breed nowhere else) a space is cleared and sticks and a few briars are collected there sufficient for them to avoid attacks of Accipitres and Aquilæ. They lay their eggs and sit; so soon as these are hatched, they lead their young away forthwith because they cannot supply food to them by means of flight. Both Coturnices and Perdices cherish their chicks beneath them, themselves leading them in the same way as hens lead theirs. They do not lay and incubate in the same place (year after year), lest any one should find it through the length of time for which they sit. Should

[1] See p. 35.

Cùm ad nidum quis uenando acceſſerit, pro-
[p. 114] uoluit ſe perdix ante pedes uenantis, quaſi iam capi poſſit[1], atque ita ad ſe capiendam hominem allicit, eouſque dum pulli effugiant, tum ipſa uolat, & reuocat prolem. parit oua non pauciora quàm decem.

Eſt & alia auis, quæ perdix ruſtica dicitur, Anglis rala dicta, cuius his uerſibus Martialis[2] meminit:

Ruſtica ſum perdix, quid refert ſi ſapor idem?
Charior eſt perdix, ſi ſapit illa minus.

DE PHASIANO.

Phaſianus, Anglicè a pheſan, Germanicè eyn faſant/ oder eyn faſian.

PLINIUS[3].

Phaſiani geminas aures ex pluma ſubmittunt, ſubriguntq;. quæ ueluti cornicula apparent.

ARISTOTELES[4].

Phaſianorum oua punctis diſtincta ſunt ut meleagridum, puluerant ut gallinæ & perdices.
[p. 115] Phaſiani à pediculis infeſtantur, & niſi interdum puluerent, eiſdem interimuntur.

DE PHŒNICE.

PLINIUS[5].

Aethiopes atque Indi diſcolores maximè & inenarrabiles ferunt aues, & ante omnes nobilem Arabię phœnicem: haud ſcio an fabuloſè, unum in toto orbe, nec uiſum magnoperè. Aquilæ narratur magnitudine, auri fulgore circa colla, cætera purpureus, cœruleam roſeis caudam pen-

1 Aristotle has ὡς ἐπίληπτος οὖσα, which means 'as if disabled.'
2 *Epigr.* Lib. XIII. lxxvi.
3 *Hist. Nat.* Lib. X. cap. xlviii.
4 *Hist. An.* Bk VI. 5, Bk IX. 260, Bk V. 140.
5 *Hist. Nat.* Lib. X. cap. ii.

any man when hunting come up to the nest, the Perdix casts herself before the hunter's feet, as if she could be caught at once, and thus attracts the man to take her, till the chicks escape, whereon she flies off and recalls her brood. She lays no fewer than ten eggs.

And there is yet another bird, which is called Perdix rustica: it is called "rale" by Englishmen, and of it Martial makes mention in the following lines:—

I am a country partridge, but what matters it so that the flavour be the same? The partridge is the dearer, if it tastes less well[1].

Of the Phasianus.

Phasianus, in English a phesan, in German eyn fasant or eyn fasian.

Pliny.

The Phasiani lower and erect two ears of feathers, which look like small horns.

Aristotle.

Of Phasiani the eggs are marked with spots, like those of Meleagrides; they dust themselves, just as Gallinæ and Perdices do. Phasiani are a prey to lice, and if they do not sometimes dust themselves, are killed by them.

Of the Phœnix.

Pliny.

The Æthiopians and Indians tell of birds of very varied colouring and indescribable, and of the Phœnix of Arabia, most noteworthy of all: I know not whether falsely, that there is but one in the whole world, and this not often seen. It is declared to be of the size of an Aquila, with golden sheen around the neck, but purple otherwise, varied with roseate feathers on

[1] The text here given is probably corrupt, which makes the rendering uncertain.

nis diſtinguentibus, criſtis faciem caputq́; plumeo apice honeſtante. Primus atque diligentiſſimus togatorum de eo prodidit Manilius, ſenator ille maximus, nobilis, doctore nullo, autor eſt, neminem extitiſſe, qui uiderit ueſcē-[p. 116]tem. Sacrum in Arabia ſoli eſſe, uiuere annis 660. ſeneſcētem caſia thurisq́; ſurculis conſtruere nidum, replere odoribus, & ſuperemori. Ex oſſibus deinde ac medullis naſci primò ceu uermiculum, inde fieri pullum.

DE PICA.

Κίττα, *pica, Anglicè a py, or a piot, Germanicè* eyn elſter/ oder eyn atzel.

ARISTOTELES[1].

Pica uoces plurimas commutat, ſingulis enim ferè diebus diuerſam emittit uocem. Parit oua circiter nouem numero. Nidum in arboribus facit ex pilis & lana, glandes cùm deficiunt, colligit, & in repoſitorio abditas, reſeruat.

PLINIUS[2].

Minor nobilitas, quia non ex longinquo [p. 117] uenit, ſed expreſſior loquacitas certo generi picarum eſt, quàm pſitacis eſt. Nec diſcunt tantùm, ſed diligunt meditantesq́; intrà ſemet, cura atq; cogitatione, intentionem non occultant. Conſtat emori uictas difficultate uerbi, ac niſi ſubinde eadem audiant, memoria falli, quærentes mirum in modum hilarari, ſi interim audierint id uerbum. Nec uulgaris ijs forma, quamuis non

[1] *Hist. An.* Bk IX. 81.
[2] *Hist. Nat.* Lib. X. cap. xlii.

a tail of blue, tufts beautifying the face, a feathery crown the head. First of our citizens and with great care Manilius wrote of it, that noted senator, of such high birth ; of his own knowledge he asserts that nobody exists who ever saw it eat. He says that in Arabia it is considered sacred to the sun, and lives for six hundred and sixty years. When it grows old it makes itself a nest with cassia and twigs of frankincense, this nest it stores with scents and on the top it dies. Then from its bones and marrow is produced what seems a little worm, but afterwards becomes a chick.

Of the Pica.

Κίττα, pica, in English a py or a piot, in German eyn elster or eyn atzel.

Aristotle.

The Pica oftentimes changes its notes, for almost every day it utters different cries. It lays about nine eggs. It makes a nest in trees, of hair and wool, and when acorns grow scarce, it gathers them and keeps them hidden in store.

Pliny.

Less fame, because it does not come from distant lands, though more distinct loquacity characterizes a certain sort of Picæ than the Psittaci. Not only do they learn, but they delight to talk, and meditating carefully and thoughtfully within themselves hide not their earnestness. They are known to have died when overcome by difficulty in a word, and, should they not hear the same things constantly, to have failed in their memory, and while recalling them to be cheered up in wondrous wise, if meanwhile they have heard that word. Nor[1] is their beauty of an

[1] Or perhaps, 'their form is not commonplace, though not showy to the eye.'

ſpectanda, ſatis illis decoris in ſpecie ſermonis humani eſt. Verùm addiſcere alias negant poſſe, quàm quæ ex genere earum ſunt, quæ glande ueſcantur, & inter eas faciliùs, quibus quini ſunt digiti in pedibus: ac ne eas quidem ipſas, niſi primis duobus uitę annis. Nuper et adhuc tamen rara ab Appennino ad urbem [p. 118] uerſus cerni cœpere picarum genera, quæ longa inſignes cauda uariæ appellantur, proprium his calueſcere omnibus annis cùm feratur rapa.

Plinius duo picarum genera facere uidetur: poſterius hoc genus Plinij, picarum genus eſſe uidetur, quod paſſim in Germania & Anglia longa cauda præditum, oua & pullos gallinarum populatur. Aliud genus picæ, tam longa cauda ornatum, quàm hoc eſt, non noui. noſtra quoque pica uulgaris calueſcere quotannis ſolet. Alterum autem picæ genus diu ſanè dubitaui quod nam eſſet, & adhuc non ſatis teneo. Cùm eſſem in Italia ad ripam Padi, ambulantibus mihi, & itineris mei comitibus, auis quædam picæ ſimilis, lingua Britannica iaia, & Germanica mercolphus appellata, conſpiciendam ſeſe commodùm obtulit, cuius nomen Italicum quum à monacho quodam, qui tum fortè aderat, percontarer, picam granatam dici reſpondit. Qua re cùm apud Italicum etiam uulgus non ſolùm priſtinæ linguæ Romanæ, ſed & rerum ſcientiæ, non obſcura ueſtigia adhuc ſupereſſe depræhenderem, ſuborta eſt mihi hinc ſuſpicio, auem hanc è generibus picarum eſſe, & quòd [p. 119] *ſcirem eandem, altera uulgari pica, multò expreſſius humanas uoces imitari, ita ſuſpicionem meam auxit, ut parùm abſit, quin credam hanc eſſe alterius generis picam, nam & glandibus ueſcitur magis omnibus alijs auibus.*

Iaia Anglorum, mercolphus Germanorū.

Hanc meā opinionē Perottus[1] cōfirmat, quem ſi uacat lege.

[1] Probably Reader in Greek to Edward VI. (cf. *Dict. National Biogr.* XLV. p. 21).

ordinary sort, though not considerable to the eye; for them it is enough honour to have a kind of human speech. However people deny that others are able to learn, save those belonging to the group which lives on acorns—and of these again those with the greatest ease which have five toes upon each of their feet: nor even they except during the first two years of life. Of late, however, and as yet infrequently, towards the city from the Appennines there have begun to be observed some sorts of Picæ which being remarkable for the length of their tails have been called "variæ." They have this special mark that they grow bald in every year when rape is sown.

Of Picæ Pliny seems to make two kinds: this latter kind of his would seem to be that Pie which here and there in Germany and England plunders both the eggs and chicks of fowls, possessing a long tail. I do not know another kind of Pie provided with a tail so long as this. And furthermore our common Pie is wont to grow bald every year. Now what the second sort of Pie might be I doubted very long, nor have I yet grasped it sufficiently. But when I was in Italy upon the banks of the Po, and while my fellow-travellers and I were walking out, a certain bird like a Pie, in English called a Jay, in German mercolphus, offered itself conveniently for observation. Thereupon I asked a certain monk, who then by chance was present, its Italian name, and he replied to me that it was called the Seed Pie[1]. When therefore I perceived that with the common people of Italy not only patent traces of the old Roman tongue still actually existed, but also of things scientific, a suspicion rose within me that this bird was of the group of Pies; moreover, since I knew that the same imitated human tones much more correctly than the other Pie, which is the commoner, so much was I confirmed in my suspicion that I can scarcely refuse to credit that this Pie was Pliny's second kind, particularly as it lives on acorns more than any other bird.

[1] Ghiandaja is the modern Italian name, derived from 'glans'=an acorn.

DE PICO MARTIO.

a rayn byrde.

Δρυοκολάπτης, *picus martius*, *pipo*, *iynx*, *torquella*, *turbo*, *Anglicè & Germanicè a ſpecht*, eyn ſpecht.

Aristoteles[1].

Alia culicibus[2] gaudent, nec alio magìs quàm uenatu culicum uiuūt, ut pipo tum maior tum minor, utrumque picum martium uocant. Similes inter ſe ſunt, uocemq; ſimilem emittunt, ſed maiorem, quæ maior eſt. Item κόλιος[3], cui magnitudo quanta turturi ferè eſt, color luteus, lignipeta hic admodum eſt, magnaq; ex parte more picorum νέμεται ἐπὶ τῶν ξύλων, quod eſt, ut interpretatur Gaza, ex macerie uiuit: uocem emittit magnam, incola maximè Peloponeſi hic eſt.

Galgulum interpretatur Gaza.

Id eſt, uiuit ad ligna.

[p. 120]

Obſerua ubi Ariſtoteles duo tantùm picorum genera facit, ibidem illum galgalum deſcribere, & ubi tria facit, eundem omittere.

Aristoteles[4] lib. 9. cap. 9. de historia animalium.

Alauda gallinago, & coturnix nunquam in arbore conſiſtunt, ſed humi. Contrà atq; picus martius, qui nunquam humi conſiſtere patitur. Tundit hic quercus, uermium & culicū cauſa, quo exeant: recipit enim egreſſos lingua ſua, quam maiuſculam, & latiuſculam habet. Scādit per arborem omnibus modis: nam uel reſupinus, more ſtellionū, ingreditur. Vngues etiam habet commodiores quàm monedula[5], ad tutiorem arborum reptationem, his enim adfixis aſcendit. Sunt pici Martij cognomine, tria genera: unum minus quàm merula, cui rubidæ

[p. 121]

[1] *Hist. An.* Bk VIII. 43—44.

[2] Gaza translated Aristotle's σκνίψ by culex (=gnat). Most probably it may be used for various small winged creatures.

[3] There is another reading, κελεός.

[4] *Hist. An.* Bk IX. 66—69.

[5] κολοιὸς may be a misreading here and below for κελεὸς, but this seems doubtful.

OF THE PICUS MARTIUS.

Δρυοκολάπτης, picus martius, pipo, iynx, torquella, turbo, in English and in German a specht, eyn specht.

ARISTOTLE.

Some birds delight in grubs, and as a rule live on no other prey, as do the great and little Pipo, both of which people call Picus Martius. Resembling one another they utter like cries, although the greater has the louder cry. Again there is the κολιός, the size of which is, nearly as may be, that of the Turtur, and its colour yellowish. It pecks wood freely, and, as the Pici do, lives for the most part on the trunks, that is, lives on the wood[1], as Gaza renders it: it utters a loud cry, and is especially a resident in the Peloponnese.

Note that, when Aristotle only makes two sorts of Pici, in that passage he describes the Galgulus, when he makes three, he does not mention it.

ARISTOTLE BOOK 9, CHAP. 9, OF THE HISTORY OF ANIMALS.

Alauda, Gallinago, and Coturnix never alight on trees, but always on the ground. It is however otherwise with Picus Martius, which never can endure sitting upon the ground. It hammers oaks for worms and grubs, that they may shew themselves, and when they issue forth it takes them on its tongue, which it has somewhat long and broad. It climbs about a tree in every way, for it even walks upside down, after the way of Lizards. It has claws better formed for creeping safely on the trees than even the Monedula, and climbs with them stuck in. There are three sorts of birds that have the special name of Picus Martius, one less than a Merula, which has some

[1] Cf. p. 88.

aliquid plumæ ineſt. Alterũ maius quàm merula: tertium non multò minus quàm gallina. Nidulatur in arboribus tum alijs cum oleis. Paſcitur formicis & coſſis. Cùm coſſos uenatur, tam uehementer excauare, ut ſternat arbores dicitur. Iam uerò miteſcens, quidam amygdalũ, quod rimæ inſeruiſſet ligni, ut fixũ conſtanter ictum reciperet, tertio ictu pertudit, & nucleum edit. Paucis[1] quibuſdam utrinque bini ungues ut auiculæ, quam Iyngem uocant. Hæc paulò maior quàm fringilla eſt, corpore uario. Habet ſibi propriam digitorum, quã modò dixi, diſpo- [p. 122] ſitionem, & linguam ſerpentibus ſimilem: quippe quam in longitudinem, menſura quatuor digitorum porrigat, rurſumq; contrahat intra roſtrum, collum etiam porrigit in auerſum reliquo quieſcente corpore, modo ſerpentium, unde torquilla uulgò appellata eſt: quanquam turbo ab antiquis. Vngues ei grandes, & ſimiles, ut monedulis exeunt, uoce autem ſtridet.

Picus minimus. Medius. Maximus.

Primum pici genus Angli ſpechtam, & uuodſpechtam, Germani elſterſpechtam nominant. Secundum genus Angli huholam, hoc eſt, foraminum dolatorem, Germani grunſprechtã nuncupant. Tertiũ genus Anglia nõ nouit, Germani aũt craſpechtam .i. cornicinum picum appellãt, quòd cornicem plumarũ colore & magnitudine etiam penè æquet.

Plinius præter hæc tria Ariſtotelis genera, quartum pici genus facere uidetur, nam lib. 10. *cap.* 33 *ſcribit. picum aliquem ſuſpendere nidum in ſurculo primis in ramis cyathi modo, ut nulla quadrupes poſſit accedere.*

Præter uireonem ſolum, in Europa nullam aliã [p. 123] *auem ita nidulari noui. Quare nullam aliam, quàm hanc, quæ pici quartum genus eſſe poſſit, inuenio.*

[1] *Hist. An.* Bk II. 46—47.

feathers red. The second bigger than a Merula, the third not much less than a barn-door hen. It breeds in various trees and olives in particular. It feeds on ants and grubs, and when hunting for grubs is said to excavate so vigorously as to fell trees. Indeed one that was tamed broke at the third attempt an almond which it had inserted in a chink of the wood, that being fixed it might more surely receive the stroke, and ate the kernel out. In some few birds there are two claws in front and two behind, as in the little bird which men call Iynx. This kind is not much larger than a Fringilla, and has the body mottled. It has moreover the peculiar arrangement of the toes, of which I have just spoken, and a tongue like that of serpents, for it shoots it out up to a distance of four fingers' length, and draws it back again within the beak; it twists its neck moreover backwards, with its body still, just as the serpents do, whence it is commonly called Torquilla, although it is the Turbo of old writers. It has claws of great size, which are like those that grow on the Monedula, it has a strident cry.

Of Picus the first kind the English call the Specht and Wodspecht, which the Germans name the elsterspecht. The second kind Englishmen term Hewhole, that is, hewer of holes, the Germans grunspecht. The third kind England knows not, but in Germany they call it craspecht or the Crow-Picus, for it is very nearly like a Crow in colour of the plumage and also in size. Besides these three sorts of Aristotle Pliny seems to make a fourth, for in Book 10 and chapter 33 he tells us that a certain Picus hangs its nest, in fashion like a cup, upon a twig among the outer branches of a tree, so that no quadruped is able to come nigh. Except the Vireo alone, I know no other bird in Europe which places its nest in such a way. Wherefore I find no other than the above which the fourth kind of Picus possibly can be.

DE PSITACO.

Pſitacus, Anglicè a popiniay, Germanicè eyn papegay.

ARISTOTELES [1].

Nam & Indica auis, cui nomen pſitacæ, quam loqui aiunt [2], talis eſt, & loquacior [3] quum biberit uinum, redditur.

PLINIUS [4].

Super omnia humanas uoces reddunt pſitaci, & quidem ſermocinantes. India auem hanc mittit. Pſitacen uocant, uiridẽ toto corpore, torque tantũ miniato in ceruice diſtinctã. Imperatores ſalutat, & quæ accipit uerba, pronunciat: in uino p̃cipuè laſciua. Capiti eius duritia eadem quæ roſtro. Hæc cũ loqui diſcit, ferreo uerberatur radio: nõ ſentit aliter ictus. [p. 124] Cùm deuolat, roſtro ſe excipit, illi innititur, leuioremq; ſe ita pedum infirmitati facit.

DE PLATEA.

Πέλεκαν, *platea, platelea, pelecanus, Anglicè a shouelard, Germa.* eyn leffler/ od' eyn löffel gãß.

ARISTOTELES [5].

Platea fluuiatilis, conchas maiuſculas, leuesq; deuorat, quas ubi ſua ingluuie coxerit, euomit, ut hiantibus teſtis exuens, legat atque edat.

PLINIUS [6].

Platea nominatur aduolans ad eas, quæ ſe in mari mergunt, & capita illarum morſu corri-

[1] *Hist. An.* Bk VIII. 85.
[2] τὸ λεγόμενον ἀνθρωπόγλωττον.
[3] ἀκολαστότερον = reckless.
[4] *Hist. Nat.* Lib. X. cap. xli.
[5] *Hist. An.* Bk IX. 71.
[6] *Hist. Nat.* Lib. X. cap. xl.

OF THE PSITACUS.

Psitacus, in English a popinjay, in German eyn papegay.

ARISTOTLE.

An Indian bird indeed, the Psitace by name, which people say can speak, is such a one as this, and is reported as more talkative after it has drunk wine.

PLINY.

Beyond all Psitaci repeat men's words, and even talk connectedly. India sends this bird, which they call Psitace, with the whole body green marked only by a scarlet ring upon the nape. It will pronounce "Hail Emperor," and any words it hears; it is especially sportive after wine. The hardness of the head is the same as of the beak. And when the bird is being taught to speak, it is beaten with an iron rod, else it feels not the strokes. When it flies down it receives its weight upon its beak, and supports itself thereon; and thus lightens itself to remedy the weakness of its feet.

OF THE PLATEA.

Πελεκάν, platea, platelea, pelecanus, in English, a shovelard, in German eyn lefler or eyn löffel ganss.

ARISTOTLE.

The Platea, a river bird, devours biggish shell-fish, if they be but smooth, and, after it has seethed them in its crop, it casts them up again, that stripping them off from their gaping shells, it so may pick and eat them.

PLINY.

The Platea, as it is called, flies at those birds which dive below the sea, and seizes their heads with a bite

piens, donec capturam extorqueat. Eadem cùm ſe deuoratis impleuit conchis, calore uentris coctas, euomit, atque ita eſculenta legit, teſtas excernens.

[p. 125] HIERONYMUS.

Pelicani cùm ſuos à ſerpente filios occiſos inueniunt, lugent, ſeq; & ſua latera percutiunt, & ſanguine excuſſo, corpora mortuorum ſic reuiuiſcunt.

Conradus Geſtnerus, cùm Tiguri agerem, homo ut doctiſſimus, ita candidiſſimus, huius mihi auis cognitionem (ut fatear, per quem profeci) primus omniū communicauit, & ideo Germanis leflerā uocari docuit, quòd roſtrū cochleari ſimile haberet.

DE PORPHYRIONE EX PLINIO[1].

Porphyrio ſolus morſu bibit: idem ex proprio genere omnem cibū aqua ſubinde tingens, deinde pede ad roſtrum ueluti manu adferēs, laudatiſſimi in comagene[2]. Roſtra ijs & prælonga crura rubent.

DE REGVLO.

Τροχίλος, πρέσβυς, βασιλεύς, *trochilus, ſenator, regulus, Anglicè a uuren, Germanicè* eyn kuningſgen/ oder eyn zaunküningk.

[p. 126] ARISTOTELES[3].

Trochilus & fruteta incolit, & foramina, capi difficulter poteſt, fugax atque infirmis moribus eſt, ſed uictus probitate, & ingenij ſolertia præditus. uocatur idem ſenator & rex, quam ob rem aquilam cum eo pugnare referunt.

[1] *Hist. Nat.* Lib. X. cap. xlvi.

[2] It almost seems as if 'Comagene' should be 'Commageno,' in which case we might translate 'They are highly prized for ointment.'

[3] *Hist. An.* Bk IX. 75.

until it wrenches their prey from them. So too when it has filled itself with shell-fish that it has devoured, it casts them up, seethed by its belly's heat, and so picks out the eatable parts, sifting off the shells.

Hieronymus.

Pelecani, when they find their young killed by a serpent, mourn, and beat themselves upon their sides, and with the blood discharged, they thus bring back to life the bodies of the dead.

Conrad Gesner, a man most learned as he also was most truthful, first imparted to me while I was at Zurich knowledge of this bird (that I may own from whom I profited), and taught me that it was called lefler by Germans because it has a spoon-shaped beak.

Of the Porphyrio from Pliny.

The Porphyrio alone drinks with a bite, it also is peculiar in dipping all its food from time to time in water, and then bearing it to its beak with its foot, as with a hand. The best are found in Comagene. Their beaks and very long legs are red.

Of the Regulus.

Τροχίλος, πρέσβυς, βασιλεύς, trochilus, senator, regulus, in English a wren, in German eyn kuningsgen or eyn zaunküningk.

Aristotle.

The Trochilus inhabits shrubberies and holes, and cannot easily be caught. Now it is shy and of a feeble habit, but endowed with great ability of getting food and knowledge of its craft. The same is called both senator and king, on which account the Aquila, they say, fights with it.

Trochilus, eſt auium omnium minima, cauda longa & ſemper erecta, roſtro longiuſculo, ſed tenuiſſimo, colore ferè fuluo. nidum facit foris ex muſco, intus ex plumis aut lana, aut floccis, ſed plurimũ ex plumis. Oui erecti & in altero ſuo fine cõſiſtentis, formã nidus habet, in medio ueluti latere hoſtiolum eſt, per quod ingreditur & egreditur. In poſticis ædibus & ſtabilis ſtramine tectis, interdũ nidũ conſtruit, ſed ſæpius in ſyluis. auis eſt etiam ſoliuaga, & gregatim nunquam uolat, imò quoties alium ſui generis offendit, mox illi bellum indicit, & conflictatur. Quare aues illæ, quæ in Bauaria pennis auricoloribus, quas in capitibus ceu coronas aureas ferunt, in ſyluis æſtate degentes, & gregatim ad urbes hyeme aduolantes, reguli non ſunt ſed tyranni Ariſtotelis, ut poſtea docebo.

[p. 127]

DE RVBECVLA ET ruticilla.

Rubecula

Ερίθακος, ἡ ἐριθέα, *rubecula, Anglicè a robin redbreſte, Germanicè* eyn rötbruſt/ oder eyn rötkelchen.

Ruticilla.

Φοινικουρός, *&, ut alter textus habet,* φοινικούργος, *Plinio phœnicurus, Gazæ ruticilla, Anglicè a rede tale, Germanicè* eyn röt ſtertz.

Aristoteles[1].

Rubecula & ruticilla uermibus aluntur. Rubeculæ, & quæ ruticillæ appellantur, inuicem tranſeunt, eſt'que rubecula hyberni temporis, ruticilla æſtiui, nec alio ferè inter ſe differunt, niſi pectoris colore & caudę[2].

[1] *Hist. An.* Bk IX. 256.

[2] The three words 'pectoris & caudę' are not in Aristotle.

The Trochilus is smallest of all birds, with a long tail[1] always cocked up, and a bill somewhat long but very slender, it is nearly reddish-brown in colour. The nest it makes is outwardly of moss and inwardly of feathers, wool, or down, but mainly of feathers. The nest has the form of an upright egg standing on one of its ends, while in the middle of one side there is a little postern as it were, by which the bird goes in and out. It sometimes builds its nest at the back of a house or in sheds thatched with straw, but usually in woods. It also is a bird that roves alone, and never flies in flocks; nay more, so often as it meets another of its kind it forthwith declares war, and fights. Wherefore the birds with plumes of gold-colour that they wear on their heads like golden crowns, which pass the summer in Bavaria in woods, and in the winter flock to towns, are not the Reguli of Aristotle, as I presently shall prove, but the Tyranni.

OF THE RUBECULA AND THE RUTICILLA.

'Ερίθακος, ἡ ἐριθέα, rubecula, in English a robin redbreste, in German eyn rötbrust or eyn rötkelchen.

Φοινίκουρος, and, as another text has it, *φοινικούργος*, in Pliny phœnicurus, the ruticilla of Gaza, in English a rede tale, in German eyn rötstertz.

ARISTOTLE.

The Rubecula and the Ruticilla feed on worms. Rubeculæ and Ruticillæ, as the birds are called, change into one another, and what in winter is the Rubecula in summer is the Ruticilla[2], while they hardly differ from each other save in colour of the breast and tail.

[1] Turner evidently means the Wren (*Troglodytes parvulus*), but with this the 'long tail' does not agree; perhaps there is a misprint.

[2] As Sundevall remarks, Aristotle probably only meant that the Redstart was called *φοινίκουρος* in summer and *ἐρίθακος* in winter. Sundevall ascribes the misinterpretation to Gaza, whose work Turner admittedly used. This is the more likely as the section of Aristotle quoted concerns birds which change their plumage and note at different seasons.

Omnia, quæ hîc Ariſtoteles de duabus auibus iſtis conſcripſit, Plinius ex ipſo in opus ſuum tranſcripſit. Sed uterque hac in re, aucupum relatibus magìs quàm ſua experientia nixus, à ueritatis tramite longiſſimè aberrauit, nam utraque auis ſimul conſpicitur, & rubeculæ domitæ, & in caueis alitæ, eandem perpetuò formam retinent. quin & eodem tempore nidulantes, [p. 128] *ſed modis longè diuerſis ſæpiſſimè in Anglia uidi. Rubecula, quæ non ſecus æſtate quàm hyeme rubrũ habet pectus, quàm poſſit longiſſimè ab oppidis et urbibus in denſiſſimis uepretis, & fruticetis ad hunc modum nidulatur. Vbi multa querna reperit folia, aut quernis ſimilia, ad radices ueprium, aut denſiorum fruticum, inter ipſa folia nidum conſtruit: & iam conſtructum, opere ueluti topiario folijs contegit. Nec ad nidum ubiq; patet aditus, ſed una tantùm uia ad nidum itur. ea quoque parte, qua nidum ingreditur, longum ſtruit ex folijs ante hoſtium nidi ueſtibulum, cuius extremam partem paſtum exiẽs, folijs claudit. Hæc, quæ nunc ſcribo, admodum puer obſeruaui, non tamen inficias iuerim, quin aliter nidulari poſſit. Si qui alium nidulandi modum obſeruauerint, êdant, & huiuſmodi rerum ſtudioſis, & mihi cum primis nõ parùm gratificabuntur. Ego, quod uidi, alijs candidè ſum impertitus.*

Rubeculę nidulatio

Phœnicurus, quem rubicillam[1] *uocat, in excauatis arboribus & (quod ſæpè expertus ſum) in rimis & fiſſuris murorum, poſticarum ædiũ, in medijs urbibus, ſed ubi hominum minor frequentia concurſat, nidulatur. Phœnicurus mas nigro eſt capite, & cauda rubra, cætera fœminæ, niſi quòd ſubinde cantillat, ſimilis. Caudam* [p. 129] *ſemper motitat uterque. Phœnicura fœmina, & proles adeo rubeculæ pullis ſimiles ſunt, ut uix ab oculatiſſimo diſcerni poſſint. Verùm motu caudæ dignoſcuntur. Rubeculæ licet caudam moueant, poſtquam tamen ſubmiſerint, ſtatim erigunt, nec tremit bis aut ter more ruticillarũ. Ruticillæ enim ſimul atque caudam mouere ceperint, non ceſſant donec ter aut quater ſimul*

Phœnicuri nidulatio.

Rubeculã æſtate cantantẽ nunquam audiui.

[1] No doubt a misprint for Ruticilla, cf. pp. 154, 160.

All that Aristotle here has written of these two birds Pliny has copied from him into his own work. But in this matter each of them, relying on the tales of fowlers more than on his own experience, has wandered greatly from the path of truth. For both the birds are seen at the same time; moreover tame Rubeculæ, when fed in cages, constantly retain the same appearance. Moreover I have very often seen the birds in England nesting at the same time, though in very different ways. The Rubecula, which has a ruddy breast no less in summer than in winter, nests as far as possible from towns and cities in the thickest briers and shrubs after this fashion. Where it finds oak leaves in plenty, or leaves like the oak, it builds its nest among the leaves themselves close to the roots of briers or the thicker shrubs: and when completed covers it with leaves as if with topiary work. Nor does access lie open to the nest on every side, but by one way alone is entrance gained. And at that place where it enters the nest the bird builds a long porch of leaves before the doorway and, on going forth to feed, closes the end with leaves. But, what I now describe, I first observed when quite a boy, nevertheless I am not going to deny that it may build otherwise. If any have observed another way of nesting, let them tell it, and they certainly will not a little gratify the students of such things, myself among the first. I have imparted truthfully to others what I saw.

The Phœnicurus which he [Gaza] calls the Rubicilla nests in hollow trees and (as I often have had experience) in chinks and cracks of walls and outhouses in the midst of our towns, though where the throng of men is not so great. The male has a black head, a red tail, but otherwise is like the female, save that he repeatedly utters a little song. Either sex flirts the tail continually. The female Phœnicurus and its brood are so much like young of Rubecula that they can scarcely be distinguished by the sharpest eye. But by the motion of the tail they may be recognised. For the Rubeculæ, although they move the tail, yet, after they have lowered it, at once raise it again, nor does it quiver twice or thrice as does that of the Ruticilla. For no sooner have the Ruticillæ once begun to move the tail than they go on till they have lightly moved it three or four times altogether

leuiter mouerint, ut alas, iuniores auiculæ cibum à matribus efflagitantes, motitant. Rubeculæ in æſtate, ubi in ſyluis ſatis ſuperq; alimenti ſuppetit, nec ullo infeſtantur frigore, (quæ res cogit illas in hyeme ad urbes, oppida & pagos confugere) cum prole ad deſertiſſima quæq; loca ſecedunt. Quare, minùs mirandum eſt, rubeculas in æſtate non paſſim occurrere. Ruticillas quid miri eſt in hyeme nõ eſſe obuias, quum per totam hyemem deliteſcant? Adhæc cùm rubeculæ pulli, in fine autumni perfectam ferè in pectoribus rubedinem nacti, ad pagos & oppida propiùs accedunt, ruticillæ, quæ antea per totam æſtatem cernebantur, diſparent, nec amplius in proximum uſque uer cernuntur. Quæ quum ita ſe habeant, quid Ariſtoteli aut illi hoc referentibus erroris anſam præbuerit, facilè quiuis poteſt colligere.

DE RVBETRA.

[p. 130] Βάτις, *Latinè rubetra dicta, ab Ariſtotele inter auiculas uermiuoras numeratur. Porrò, quæ'nam auis ſit, prorſus diuinare non poſſum. Gybertus Longolius linariam, ſiue miliariam eſſe rubetrã putabat, quòd rubis crebrò inſideat. Sed quum Anglorum buntinga in rubis tam frequẽs ſit, quid uetat quô minùs & ipſa quoq; batis dici poſſit. Nihil igitur certi habemus, quod nomen Britannicum aut Germanicum ſit huic aui imponendum.*

Sed quum auium ſuprà commemoratarum altera ſeminibus herbarũ ueſcatur, & altera hordeo & tritico, & batis Ariſt. uermiuora ſit, delegẽda eſt auicula quæpiam, quæ ſolis uermibus paſcitur, qualis eſt auicula Anglis ſtonchattera, aut mortettera dicta, & Germanorum klein brachuogelchen. Hæc ſi batis non ſit, mihi prorſus ignota eſt. Porrò illa, quam Angli linotam, & Germani flasfincam uocant, ueteribus, ſi Ruellio, credimus, erit miliaria.

just as young small birds flutter their wings soliciting meat from their mothers. In summer, when there is enough and more of food found in the woods, and they are not troubled by any cold (a thing which forces them in winter to resort to cities, towns, and villages), Rubeculæ retire to the most solitary places with their young. And so it is no marvel that Rubeculæ do not occur in summer everywhere. And what wonder is it that Ruticillæ are not met with in winter, since throughout the whole of winter they are hidden? And further, when the young Rubeculæ, having almost assumed the full red on their breasts at the end of autumn, come nearer to towns and villages, the Ruticillæ, which were hitherto seen during the whole summer, disappear and then are no more noticed till the following spring. Wherefore, things being thus, anyone may easily perceive what gave a handle to Aristotle or to those who reported this error to him.

OF THE RUBETRA.

Βατίς, in Latin called Rubetra, is by Aristotle classed among the little birds that feed on worms. Beyond this I cannot guess at all what sort the bird may be. Yet Gybertus Longolius[1] believed that the Rubetra was the Linaria or the Miliaria, because it often perched on brambles. But since the Bunting of the English sits so commonly on brambles, what forbids that bird from also being called the Batis? On this account we have no certainty as to what name, British or German, should be given to this bird.

But inasmuch as of the birds mentioned above the one eats seeds of grasses, and the other wheat and barley, and as Aristotle's Batis lives on worms, some small bird must be chosen which eats worms and nothing else. Now such a little bird is that called by the English Stonchatter or Mortetter and the klein brachvogelchen of the Germans. If this be not the Batis, it is quite unknown to me. Besides that which the English call the Linot and the Germans the flasfinc must be the Miliaria of older works, if we believe Ruellius[2].

[1] For this author see Introduction.

[2] Ruellius wrote *De natura stirpium libri tres* (1536) and edited one or more medical or other works.

DE RVBICILLA.

Πυῤῥούλας, *rubicilla, Angli. a bulſinche, Germa.* eyn blödtfinck.

Rubicillam Ariſtoteles inter eas aues connumerat, quæ uermibus ueſcuntur: ſed pluribus uerbis eam non deſcribit. Ego nominis etymologiam ſecutus, rubicillam [p. 131] *Anglorum bulſincam, & Germanorum blouduincam eſſe conijcio. Nam omnium, quas unquam uidi auium mas in hoc genere, pectore eſt longè rubidiſſimo: fœmina uerò pectore toto eſt cinereo, cætera mari ſimilis. Sed ut faciliùs omnes intelligant, de qua aue ſcribam, magnitudine paſſeris eſt, roſtro breuiſſimo, latiſſimo, et nigerrimo, lingua latiore multò quã pro corporis magnitudine. Pars ea linguæ, quæ cibi ſapores dijudicans, oris cœlum tangit, carnea & nuda eſt, reliquæ partes cornea pellicula obducuntur. Supremam auis partem plumæ cyaneæ contegunt. cauda nigra eſt, & capite etiam nigro, ueſcitur libentiſſimè primis illis gemmis ex arboribus ante folia & flores erumpentibus, & ſemine canabino. auis eſt imprimis docilis, & fiſtulam uoce ſua proximè imitatur. nidulatur in ſepibus, & oua quatuor excludit, ut plurima quinq;. eundem colorem per totum annum ſeruat, nec locum mutat. Quæ quum ita ſe habeant, non poteſt hæc atricapilla eſſe, ut quidam uolunt, utcunq; extremo linguæ acumine carere uideatur.*

DE SALO, QVI GRÆCE ἄιγιθος *dicitur.*

ARISTOTELES [1].

[p. 132] Salus uitæ commoditate, & partus numero commendatur, ſed alterius pedis clauditate cedit. Sali & flori ſanguinẽ miſceri negant, tã ingens inter ſalum & florũ feruet odiũ. Salo etiam preliũ cũ aſino eſt, propterea quòd aſinus ſpi-

[1] *Hist. An.* Bk IX. 89, 22, 14.

Of the Rubicilla.

Πυῤῥούλας[1], rubicilla, in English a bulfinche, in German eyn blödtfinck.

Aristotle counts the Rubicilla among those birds which feed on worms: but he does not describe it in more words. I, following the derivation of the name, conjecture that it is the Bulfinc of the English and the bloudvinc of the Germans. For of all the birds I ever saw the male of this kind has by far the reddest breast: the female however has the breast wholly grey, though otherwise like the male. But, that all may understand more easily about which bird I write, it is the size of a Sparrow, with the beak particularly short and broad and black, the tongue much broader than is in proportion to its body. That part of the tongue which discriminates the flavour of the food and meets the palate of the mouth is flesh-coloured and naked, while the other parts are covered with a horny pellicle. Bluish grey feathers clothe the upper parts. The tail is black and the head also black. It feeds most greedily on those earliest buds, which burst out on the trees before the leaves and flowers, as well as hemp-seed. It is the readiest bird to learn, and imitates a pipe very closely with its voice. It nests in hedges where it lays four eggs or five at most. It keeps the same colour throughout the year, and does not change its home. And since these things are so, it cannot be the Atricapilla, as some will have it, though it may appear to lack the point at the tip of the tongue.

Of the Salus, which in Greek is called αἴγιθος.

Aristotle.

The Salus is well thought of for its skill in gaining a livelihood, and for the number of its young, although it suffers from lameness in one of its feet. And men deny that the blood of the Salus and the Florus ever mixes, for so great an enmity rages between the birds. There is war also between the Salus and the Ass,

[1] Sundevall thinks that the Πυῤῥούλας is the Redbreast, but the description does not agree with his idea. Another reading is πυρρουράς.

netis ſua ulcera ſcabendi cauſa atterat: tum igitur ob eam rem, tũ etiam quòd ſi uocem rudentis audierit[1], oua abigat per abortum, pulli etiam metu in terram labantur. Itaque ob eam iniuriam aduolãs, ulcera eius roſtro excauat.

PLINIUS[2].

Aegithus auis minima cum aſino pugnat, ſpinetis enim ſe ſcabendi cauſa atterens, nidos eius diſſipat, quod adeò pauet, ut uoce audita omnino rudentis oua eijciat, & pulli ipſi metu [p. 133] cadant: igitur aduolans ulcera eius roſtro excauat.

DE SITTA.

Sitta, Anglicè a nut iobber, Germanicè eyn nusshäkker oder eyn meyſpecht.

ARISTOTELES[3].

Sunt & ei, quæ ſitta dicitur, mores pugnaces, ſed animus hilaris, cõcinnus, compos uitæ facilioris. Rẽ maleficam illi tribuunt, quia rerum callat cognitione, prolem numeroſam facilemq; progignit, carioſa ligna contundens, ex coſſis, quos inde eruit, uictitat. Sitta[4] aquilæ oua frangit, aquila tum ob eam rem, quum etiam quòd carniuora eſt, aduerſatur.

Auicula, quam Angli nucipetam uocant, & Germani meyſpechtum, paro maximo paulò maior eſt, pennis cyaneis, roſtro longiuſculo, & per arbores eodem modo, quo picus aſcendit, & eaſdem uictus gratia contundit: nuces roſtro etiam perforat, & nucleos commedit. nidu- [p. 134] *latur in cauis arboribus more pici, uoce ualde acuta & ſonora eſt.*

[1] Aristotle has 'κἂν ὀγκήσηται, ἐκβάλλει τὰ ᾠὰ καὶ τοὺς νεοττούς,' as if the bray of the ass shook the eggs and young out of the nest. Gaza seems to have had the reading ἐκτίκτουσι for ἐκπίπτουσι

[2] *Hist. Nat.* Lib. X. cap. lxxiv.

[3] *Hist. An.* Bk IX. 91, very freely rendered.

[4] *Hist. An.* Bk IX. 17

because the Ass is wont to rub its sores against the thorn-bushes to scratch them, therefore for this cause, and also because the bird has heard the brayer's voice it prematurely drops its eggs, while even nestlings fall down to the ground with fear. So for that injury (the bird) attacking it scoops out its sores.

PLINY.

The Ægithus, a very little bird, wages war with the Ass, because it, rubbing against thorn-bushes to scratch itself, destroys the nest, and this the bird dreads so much that, if it merely hears the brayer's voice, it drops its eggs, and the young also fall to the ground with fear. Accordingly attacking it the bird scoops out its sores.

OF THE SITTA.

Sitta, in English a nut jobber, in German eyn nushäkker or eyn meyspecht.

ARISTOTLE.

That bird which is called Sitta has pugnacious habits but a cheerful disposition ; it is elegant and well adapted to get food with ease. Yet men attribute witchcraft to it, since it is cunning in knowledge of affairs ; it produces numerous young with ease ; hammering on rotten trees, it lives upon the grubs which thence it digs. The Sitta breaks the eggs of the Aquila, on which account, and also since it is carnivorous, the Aquila is its enemy.

The small bird which the English call the Nut-seeker and Germans the meyspecht is somewhat bigger than the biggest Parus, with blue plumage and a longish beak. It climbs trees in the same way as the Picus, and hammers the same for food ; it also bores nuts with its beak, and eats the kernels. It nests in hollow trees, as does the Picus, while its note is very sharp and loud.

DE STRVTHIONE.

Στρουθὸς, λιβυκός, *ſtruthio, aut ſtruthiocamelus, Anglicè an oiſtris, Germanicè* eyn ſtrauß.

ARISTOTELES [1].

Struthio, etiam libicus, eodem modo partim auem, partim quadrupedem refert, quippe qui non ut quadrupes pennas habeat, ut non auis ſublimis non uolet, nec pennas ad uolandum commodas gerit, ſed pilis ſimiles. Itē quaſi quadrupes ſit, pilos habet palbebrę ſuperioris, & gibber[2] capite, parte colli ſuperiore eſt. Itaq; cilia habet piloſiora, ſed quaſi auis ſit, infrà pennis integitur. Bipes etiam tanquam auis, biſſulcus tanquam quadrupes eſt. Nō enim digitos habet, ſed ungulam bipartitam. quarum rerum [p. 135] cauſa eſt, quòd magnitudine non auis, ſed quadrupes eſt. Magnitudinem enim auiū minimam eſſe propè dixerim, neceſſe eſt. Corpus enim molem ſublimem mouere, nequaquam facilè eſt.

DE STVRNO.

Ψάρος, *ſturnus, Anglicè a ſterlyng, Germanicè* eyn ſtär/ oder eyn ſtör.

ARISTOTELES [3].

Sturnus niger eſt, albis maculis diſtinctus, magnitudine merulæ.

PLINIUS [4].

Sturnorum generi proprium, cateruatim uolare, & quodam pilæ orbe circumagi, omnibus in medium agmen tendentibus.

1 *De partibus Animalium*, IV. 14.
2 No doubt a misprint for 'glaber.'
3 *Hist. An.* Bk IX. 102, freely rendered.
4 *Hist. Nat.* Lib. X, cap. xxiv.

OF THE STRUTHIO.

Στρουθός, λιβυκός, struthio or struthiocamelus, in English an oistris, in German eyn strauss.

ARISTOTLE.

The Struthio, or Libyan bird, in like manner partly recalls a bird, partly a quadruped, seeing that it, unlike a quadruped, has wings, and yet, unlike a bird, it does not fly aloft, nor has it feathers fit for flight, since they resemble hairs. Likewise as if it were a quadruped, it has hairs on the upper eyelid, while the head and upper portion of the neck are bare. So also it has somewhat hairy eyelashes, yet it is covered with feathers beneath, as if it were a bird. Moreover it is biped like a bird, but yet it is cloven-footed like a quadruped. That is, it has not toes but a divided hoof. The cause of these things is that in its size it is not a bird but a quadruped. For I would almost say that a bird's size must be extremely small, for it is by no means easy to move aloft a body when the mass is vast.

OF THE STURNUS.

Ψάρος, sturnus, in English a sterlyng, in German eyn stär or eyn stör.

ARISTOTLE.

The Sturnus is black, varied with white spots, and of the bigness of a Merula.

PLINY.

It is peculiar to Starlings in their kind to fly in crowds, and wheel about as it were in a ball, all tending to the middle of the band.

DE TINVNCVLO.

Κέγχρις, *tinunculus, Anglicè a kiſtrel, or a kaſtrel, or a ſteingall.*

ARISTOTELES[1].

[p. 136] Omnes, quibus ungues adunci, parcius generant, excepto tinunculo, qui plurima in adunco genere parit. Iam enim quatuor eius reperti ſunt pulli, ſed plures etiam procreari poſſe, apertum eſt. Tinunculo[2] uentriculus ingluuiei ſimilis eſt, & ſolus in adunco genere bibit. Rubra ſunt eius oua modo minij.

Tinunculus colore multò magis eſt fuluo quàm reliqui accipitres, & corpore paruo. Auiculas inſequitur, &, ut quidam mihi retulêre, papiliones interdum. In cauis nidulatur arboribus, & in templorum muris, & æditioribus turribus, ut apud Germanos Argentorati & Coloniæ, & apud Anglos Morpeti obſeruaui. pullos etiam diu uolantes tantiſper alit, dũ ipſi ex proprio uenatu uiuere poſſũt.

DE TETRAONE.

Τέτριξ, ὄυραξ, *tetrao, Anglicè a buſtard, or a biſtard, Germanicè* eyn träp/ oder eyn trap gänß.

ARISTOTELES[3].

[p. 137] Tetrix, quam Athenienſes uragem uocant, nec terræ, nec arbori ſuum nidum committit, ſed frutici[4].

PLINIUS[5].

Decet tetraones ſuus nitor, abſolutaq; nigritia, in ſupercilijs cocci rubor. Alterum eorum genus uulturum magnitudinẽ excedit, quorum

1 *Hist. An.* Bk VI. 2.
2 *Hist. An.* Bk II. 88, Bk VIII. 50, Bk VI. 6.
3 *Hist. An.* Bk. VI. 4.
4 Aristotle has χαμαιζήλοις φυτοῖς.
5 *Hist. Nat.* Lib. X. cap. xxii.

OF THE TINNUNCULUS.

Κεγχρίς, tinnunculus, in English a kistrel or a kastrel, or a steingall.

ARISTOTLE.

All birds with crooked claws[1] breed somewhat sparingly, save the Tinnunculus, and it of all the crooked-claw kind lays the most eggs. For of this bird four young have been already found, while it is evident that more might be produced. The stomach in Tinnunculus is not unlike a crop, whereas it is the only one of all the crooked-claw kind that drinks. Its eggs are red—like scarlet.

The Tinnunculus is of a much more fulvous colour than are other Hawks, and small in body. It chases little birds, and, as some men have told me, butterflies at times. It nests in hollow trees, church walls, and lofty towers, as I have seen in Germany at Strassburg and at Cullen, and in England at Morpeth. It also feeds its young long after they can fly until such time as they can live apart by hunting for themselves.

OF THE TETRAO.

Τέτριξ, *οὗραξ*, tetrao, in English a bustard or a bistard, in German eyn träp or eyn trap gänss.

ARISTOTLE.

The Tetrix, which Athenians call the Urax, trusts its nest not to the ground, nor to a tree, but to low-growing plants.

PLINY.

Their glossy plumage well becomes the Tetraones, as does furthermore their perfect blackness and the scarlet redness of their eyebrows. But one kind exceeds in size the vultures and recalls their colour-

[1] For Aristotle's groups of Birds see p. 35.

et colorem reddit. Nec ulla auis, excepto ſtruthiocamelo, maius corpore implẽs pondus, in tantum aucta, ut in terra quoq; præhendatur. gignunt eos alpes, & ſeptentrionalis regio. In uiuarijs ſaporem perdunt. Moriuntur contumacia ſpiritu reuocato. Proximæ eis ſunt, quas Hiſpani aues tardas appellãt, Græcia otidas dãnatas cibis. Emiſſa enim oſſibus medulla, odoris tędium extemplo ſequitur.

[p. 138]

DE TYRANNO.

Tyrannus, Anglicè a nyn murder, Germanicè eyn neun mürder/ oder eyn gold hendlin.

ARISTOTELES [1].

Veſcitur & uermibus tyrannus, cui corpus non multò amplius quã locuſtæ, criſta rutila ex pluma elatiuſcula, & cætera elegans, cantuq; ſuauis hæc auicula eſt.

Quanquam Ariſtoteles unum tantùm tyranni genus faciat, Colonienſes tamen aucupes tria genera eſſe contendunt. Primum uocant die groſſe neun murder, quod Angli etiam ſchricum nominant: & ego Ariſtoteles mollicipitem eſſe conijcio: ſturnum magnitudine æquat, color eius à cyaneo ad cinereum uergit. Secundum genus eiuſdem eſt coloris, cuius & ſuperius, ſed paſſerem magnitudine non excedit. Hoc genus etiam in aues ſæuit. Tertium genus, quod Ariſtotelis tyrannus eſt, auicula eſt regulo paulò maior, criſta rutila redimita, & cæteris generibus (ſi aucupibus credere phas[2] ſit) cede[3] & corporis effigie non diſſimilis. Secundum & tertium tyranni genus apud Anglos hactenus nunquam uidere [p. 139] *contigit, & primum genus licet in Anglia ſit, pauciſſimis tamen notum eſt: ſunt tamen, qui norunt, & shricum uocant.*

[1] *Hist. An.* Bk VIII. 41.

[2] Of course a misprint for 'fas.'

[3] No doubt this should be 'sede.'

ing. There is no other bird, except the Struthiocamelus, which attains so great a weight of body, growing to such a size that it may even be caught upon the ground. The Alps produce them, as do northern lands. In mews they lose their flavour. They die of stubbornness by holding back their breath. Very near them are those which Spaniards call "Aves tardæ" and Greece "Otides"; they are condemned as food. For when the marrow issues from the bones, disgust at the smell follows there and then[1].

Of the Tyrannus.

Tyrannus, in English a nyn murder, in German eyn neun mürder or eyn gold hendlin.

Aristotle.

The Tyrannus also lives on worms[2]. Its body is but little larger than a locust's is. This little bird moreover has a somewhat upright crest of reddish feathers, and is otherwise pretty; its song is sweet.

Though Aristotle makes but one kind of Tyrannus, yet the bird-catchers of Cullen state that there are three. The first they call die grosse neun mürder, which the English name the Schric for their part, this I take to be the Molliceps of Aristotle. In size it equals the Sturnus, while its colour verges from blue to grey. The second kind is of the same colour as the foregoing, but in size does not exceed a Sparrow. Furthermore this kind is cruel towards other birds. The third kind, which is Aristotle's Tyrannus, is a small bird which is little bigger than the Regulus, adorned with a red crest and not unlike the other kinds (if it be right to trust the fowlers) in its haunts and form of body. It has not ever happened to me hitherto in England to observe the second or the third kind of Tyrannus and, although the first exists in England, it is known to very few. Yet there are some who know it and who call it Shric.

[1] See p. 106. [2] See p. 35.

DE TRYNGA.

Τρύγγας, *trynga, Anglicè a uuater hen, or a mot hen, Germanicè* eyn wasser hen.

ARISTOTELES[1].

Lacus & fluuios petunt iunco, cinclus, & trynga[2], quæ inter hæc minora, maiuſcula eſt, turdo enim æquiparatur: omnibus his cauda motitat.

Iam, ut ſciatis, quam auem tryngam eſſe putem, auis tota pulla eſt, excepta ea caudæ parte, quæ podicem tegit, ea enim candida eſt, & tum cernitur, cùm caudam erigit. alis parùm ualet, atque ideo breues facit uolatus. In ſtagnis, quæ nobilium ædes obducunt, & in piſcinis apud Anglos plerunq; degit. Si quando periclitatur, ad arundineta denſiora ſolet confugere.

DE TVRDO.

Κίχλα, *turdus, Anglicè à thruſche, Germanicè* eyn krammeſuögel/ oder eyn wachholteruögel.

[p. 140] ARISTOTELES[3].

Turdorum tria ſunt genera, unũ uisciuorũ, quod niſi uiſco reſinaq; non ueſcitur, & magnitudine picæ eſt. Alterum pilare, quod ſonat acute. & magnitudine merulæ eſt. Tertium quod iliacum quidam uocãt, minimum inter hęc, minus'que maculis diſtinctum eſt. Mutat[4] & turdus colorẽ, quippe collo æſtate uarius, hyeme diſtinctus ſpectetur, uoce tamẽ eadẽ eſt. Turdus[5] nidos ex luto, ut hirundines, facit, in excelſis arboribus, ita deinceps continuato opere,

[1] *Hist. An.* Bk VIII. 47.

[2] Another reading is πύγαργος; but this word is elsewhere used of an Eagle, cf. p. 30.

[3] *Hist. An.* Bk IX. 96.

[4] *Hist. An.* Bk IX. 254.

[5] *Hist. An.* Bk VI. 3.

Of the Trynga.

Τρύγγας, trynga, in English a water hen or a mot hen, in German eyn wasser hen.

Aristotle.

The Junco and the Cinclus live on lakes and streams, as does the Trynga, which among these little birds is somewhat largest, for it equals in its size a Turdus; all these wag their tails.

And now, that you may know what bird I think the Trynga is, it is an altogether dusky bird, save that part of the tail which lies above the vent, for that is white and only visible when it erects its tail. It is weak on the wing, and therefore takes short flights. In England for the most part it haunts moats which surround the houses of the great, and fishponds. If danger ever threatens it is wont to flee to the thicker reed-beds.

Of the Turdus.

Κίχλα, turdus, in English a thrusche, in German eyn krammesvögel or eyn wachholtervögel.

Aristotle.

There are three kinds of Turdi, one of which is called the Viscivorus, since it feeds on naught but mistletoe and gum, and is of the size of a Pica. The second, the Pilaris, which has a sharp note, is of the same size as a Merula. The third, which some call the Iliacus, is the least of them and less marked with spots. The Turdus changes colour also, since it may be seen mottled upon the neck in summer, while in winter it is spotted, though its voice continues similar. The Turdus makes its nests of mud, as do Hirundines, and places them in lofty trees, the building

ut quaſi catena quædã nidorum contexta uideatur.

PLINIUS[1].

Turdi hyeme maximè in Germania cernuntur.

[p. 141] Turdus primus. Secundus. Tertius.

Primum turdi genus Angli peculiariter nominant a thrushe, & Germani (niſi me fallant aucupes, qui me ſic uocare docuerunt) eyn crammeſuogel. Secundum genus Angli uocant a throſſel, aut a mauis, Germani eyn droſſel, aut eyn durſtel. Tertiũ genus ab Anglis a uuyngthrushe, & à Germanis eyn uueingaerdsuoegel nuncupatur. Hic turdus utrinque iuxta oculos, & in pectore & in ipſo alæ flexu, intus & foris maculas habet latiuſculas rubras. Huius nidum nunquam uidi: nec mirum, quum per æſtatem apud nos nuſquam uideatur. primum genus non niſi hyeme in Anglia cernitur, aut ſi uideatur, rarum eſt. Secundum genus per totum annum apparet maculoſo ualde pectore, & cãtus ſui gratia à multis in caueis alitur. Nidum intus ex luto aut lignorum carie liquore mixta, & artificioſè leuigata, foris ex muſco in ramis arborum aut fruticum facit.

DE VIREONE.

Χλωρίον, *uireo, Anglicè a uuituuol, Germanicè* eyn witwol/ oder eyn weidwail/ oder eyn kerſenriſe.

ARISTOTELES[2].

Vireo docilis, & ad uitæ munia ingenioſus [p. 142] notatur, ſed malè uolat, nec grati coloris eſt. Vireo[3] totus uiridis ex obſcuro[4] eſt, hyeme hic non uidetur, ſed æſtiuo ſolſtitio potiſſimùm uenit in conſpectum. Diſcedit exortu arcturi ſyderis, magnitudine turturis eſt.

[1] *Hist. Nat.* Lib. X. cap. xxiv.
[2] *Hist. An.* Bk IX. 89.
[3] *Hist. An.* Bk IX. 98.
[4] These two words are not in Aristotle.

being so continuous as to seem almost like a chain of nests woven together.

Pliny.

The Turdi are in winter chiefly seen in Germany.

The first kind of Turdus Englishmen particularly name a Thrushe, and Germans (if the bird-catchers, who taught me so to call it, lead me not astray) eyn crammesvogel. The second kind the English call a Throssel or a Mavis, but the Germans say eyn drossel or eyn durstel. The third is named a Wyngthrushe by the English and eyn weingaerdsvogel by the Germans. This Turdus has broadish red spots on each side near the eyes, as well as on the breast, and also both inside and outside at the bend of the wing. But I have never seen its nest, nor is that wonderful, since it is nowhere to be seen with us throughout the summer. The first kind is not observed in England save in winter, or, if it be seen, it is unusual. The second kind, with a much spotted breast appears throughout the year, and by many is kept in cages for its song. It builds a nest, moreover, inwardly of mud or else of rotten wood tempered with moisture and smoothed skilfully, and outwardly of moss, upon the boughs of trees or shrubs.

Of the Vireo.

Χλωρίων, vireo, in English a witwol, in German eyn witwol or eyn weidwail, or eyn kersenrife.

Aristotle.

The Vireo is teachable, and is remarkable for its capacity for the duties of life; but it flies badly and is not of a pleasing colour[1]. The Vireo is wholly of a dusky green; it is not seen in winter here, but comes chiefly in view about the summer solstice, it departs at the rise of the star Arcturus, and is of the size of the Turtur.

[1] Cf. p. 86.

Vireonem (quod ſcio) in Anglia nunquam uidi, ſed in Germania ſæpiſſimè. turture paulò minor eſt. Vocem fiſtulæ grandiuſculæ, quæ infimam cantionis partem ſuſtinet, ſimilem emittit. Nidum in ramo quem in ſumma arbore ſuſpendit, in formam rotundam conſtruit, ne cui hominum aut ferarũ ad eum pateret aditus.

DE VPVPA.

ἔποπς, upupa, Anglicè a houupe, Germanicè eyn houp/ oder eyn widhopff.

Aristoteles[1].

Vpupa potiſſimùm nidum è ſtercore hominis facit. Mutat faciẽ tempore æſtatis & hyemis, ſicut & cæterarum quoque agreſtium plurimæ. [p. 143] Vpupa[2] una in ſuo genere non nidificat, ſed ſtipites arborum ſubiens, parit ſine ullo ſtramento, in cauis.

Anglorũ lapuuingã non eſſe upupam.

Literatores pleriq; omnes Britannici, upupam eam nominant auem, quam barbari ab alarum ſtrepitu, uannellum nuncupant, & ipſi ſua lingua lapuuingam uocant. Verùm iſtorum craſſus error facilè autoritate Plinij[3] de upupa ita ſcribentis, confutatur: Vpupa (*inquit*) obſcœna aliàs paſtu auis, criſta uiſenda plicatili, cõtrahens eam, ſubrigensq; per longitudinem capitis. *Hæc ille. Sed Grammaticis noſtris hic error eſt facilè condonandus, nam nuſquam in tota Britannia upupa (quod ego ſcio) reperiri poteſt, apud Germanos tamen frequentiſſima. Ea eſt magnitudine turdi, alis per interualla fuſcis, albis & nigris pennis diſtinctis, criſta in capite ab ea parte roſtri, qua capiti committitur, ad extremum uſq; occiput in lõgitudinem porrigitur, quam pro adfectibus ſuis aut contrahit,*

[1] *Hist. An.* Bk IX. 88.
[2] *Hist. An.* Bk VI. 4.
[3] *Hist. Nat.* Lib. X. cap. xxix.

I have never seen the Vireo in England, so far as I know, but very often when in Germany. It is a little smaller than the Turtur. It gives forth a note like that of the large pipe which plays the bass part of a song. This bird suspends its nest upon a branch at the top of a tree, and fashions it in rounded form, that it should not afford access to any man or beast.

Of the Upupa.

ἔποψ, upupa, in English a howpe, in German eyn houp or eyn widhopff.

Aristotle.

The Upupa builds its nest chiefly of human dung. It changes its appearance in the summer season and in winter, as very many other wild birds do. The Upupa only of its kind builds not a nest, but entering the trunks of trees lays eggs in cavities, without any litter.

Nearly all British writers name that bird Upupa, which from the noise of its wings foreigners call Vannellus, though in their own tongue the former call it Lapwing. Yet their gross error may be easily refuted on the authority of Pliny, who thus writes of the Upupa.

The Upupa (he says) is a bird filthy otherwise as to its food, but to be noticed for its folding crest, which it contracts and then erects again along its head.

These are his very words. And yet our scholars may be well excused this their mistake, for nowhere in the whole of Britain is the Upupa to be found (so far as I know), though in Germany it is most plentiful. The bird is of the bigness of a Thrush, with wings barred here and there with brown, and marked with black and white feathers ; the crest extends from the part of the bill which joins the head to the extremity of the occiput, along the length, this it contracts

aut dilatat, ut equus aures arrigit aut demittit. tibijs eſt ualde breuibus, alis obtuſioribus, & lentè admodum uolat.

[p. 144]

DE VRINATRICE.

Κολυμβρίς, *urinatrix, Angl. a douker, Germa.* eyn dücher.

Aristoteles[1].

Alia degunt quidem in fluido, uictumq; inde petunt, ſed aërem nõ humorẽ recipiunt, & foris párere ſolent. Complura huius generis ſunt, partim greſſilia ut lutra, latax & crocodilus: partim uolucres ut mergi & urinatrices.

Ariſtoteles urinatricis unum tantùm genus commemorat, ego tamen tria urinatricum genera uidi. Horum primum totum nigrum eſt. & ſi cirrum, quẽ in capite gerit, exceperis, mergo, quo tamen triplo minor eſt: cætera, quod ad corporis attinet effigiẽ, non diſſimile eſt. & hoc genus nautæ noſtrates lounam nominant, alij doukeram. Secundum genus turdo non maius eſt, anati colore & corporis effigie ſimile, et hoc Angli mediam urinatricem nuncupãt. Tertium genus adeo nuper ab ouo excluſum refert anſerculum, ut niſi roſtrum huius paulò tenuius eſſet, ægrè alterum ab altero diſ- [p. 145] *cerneres. Non enim pennas, ſed lanuginem quandam, earum loco obtinet. Degunt hæc plerunque tria genera in aquis ſtagnantibus, aut fluuijs non admodum rapidis, in quorum ripis arundines & carices naſcuntur.*

DE VVLTVRE.

Γύψ, *uultur, Anglicè a geir, Germanicè* eyn geyr.

Aristoteles[2].

Vultur nidificat in excelſiſſimis rupibus: unde fit ut raró nidus & pulli cernantur.

[1] *Hist. An.* Bk I. 6, somewhat freely rendered.
[2] *Hist. An.* Bk VI. 35.

or spreads again according as it is disposed, as a horse pricks or droops its ears. It has very short legs and rounded wings, while it flies somewhat slowly.

OF THE URINATRIX.

Κολυμβίς, urinatrix, in English a douker, in German eyn důcher.

ARISTOTLE.

But other animals in truth live in the water and thence seek their food, yet they breathe air and not moisture, and they are wont to breed out of the water. Now there are many of this sort, in part going afoot, as are the Lutra, Latax, and Crocodilus; and in part winged, as the Mergi and the Urinatrices.

Aristotle makes mention only of one kind of Urinatrix, but I have observed three kinds of Urinatrices. Of these the first is wholly black, and, except for the tuft it bears upon its head, is not unlike the Mergus otherwise, so far as the outline of its body goes, though it is one-third less in size. This is the sort our sailors call the Loun, but others the Douker. The second kind, no bigger than a Thrush, is like a Duck in colour and in form of body; this the English call the middle Urinatrix. The third kind, when it is but newly hatched, recalls a Gosling, so that if its beak were not a little more slender you could scarce discern the one bird from the other. For it has no quills, but in place of these a sort of down. These three kinds for the most part live on stagnant waters or not very rapid rivers, on the banks of which grow reeds and sedges.

OF THE VULTUR.

Γύψ, vultur, in English a geir, in German eyn geyr.

ARISTOTLE.

The Vulture nests in very lofty rocks, and thus it chances that the nest and young are rarely to be seen.

quocirca Herodotus Brifonis rhetoris pater, uultures ex diuerfo orbe nobis incognito aduolare putauit, argumẽto quòd nidum nemo uidiffet uulturis, & quòd multi exercitum fequẽtes, repentè appareant. Sed quanquam difficile nidum eius alitis uideris: tamen uifus aliquando eft. Pariunt uultures oua bina. Cætera, quæ carne uefcuntur, non plus quàm femel anno párere exploratum eft.

[p. 146]

PLINIUS[1].

Vulturum præualent nigri, nidos nemo attigit, ideo etiam fuêre, qui putarent ex aduerfo orbe aduolare falfo. Nidificant enim in excelfiffimis rupibus. Fœtus quidem fæpè cernuntur ferè bini. Vmbricius aurufpicũ noftro æuo peritiffimus, párere tradit tria, uno ex his reliqua oua nidũq; luftrare, moxq; abijcere. Triduo aũt antè aut biduo uolare eos, ubi cadauera futura funt.

Perperã Grammatici quidam uulturẽ, gryphem nominant, uulturem & gryphem ineptè confundentes. quum gryps fit a gryphen, animal ut creditur uolatile & quadrupes.

[1] *Hist. Nat.* Lib. x. cap. vi.

Wherefore Herodotus, the father of the rhetorician Briso, thought that Vultures winged it from some other world unknown to us, his argument being that nobody had ever seen a Vulture's nest, although a multitude at once come into sight when following an army. And yet, however difficult it be to see the nest of this bird, still it has been seen at certain times. Vultures lay two eggs each. Besides it is a well-known fact that animals which feed on flesh do not breed more than once a year.

PLINY.

Of Vultures the black are most plentiful. No one has ever reached their nests and therefore there have been some who erroneously thought that they flew hither from another world. They really nest in very lofty rocks. Indeed the offspring, generally twins, are often seen. Umbricius, the most skilful augur of our age, asserts that they lay three eggs, with one of which they cleanse the others and purify the nest, and afterwards throw it away. And that they fly three or two days beforehand to a place where carcases are likely to be found.

Quite wrongly certain scholars call the Vulture Gryps, confounding foolishly the Vulture and the Gryps, since the Gryps is a Gryphon, or an animal believed to be both winged and quadruped.

[p. 147]

AVIVM LOCI COMMVNES ex Ariſtotele.

SI Plinium, Ariſtotelẽ, Ariſtophanẽ aut quemcũq; alium idoneum ſcriptorem te legere contingat, locos huiuſmodi cõmunes, quales exempli tantũ gratia ſubijciam, in procinctu, libro inſcriptos chartaceo habere expedit, ut ad eos aues omnes, de quibus apud iſtos legis, certo referas, quod ſi feceris, nõ dubito quin in auiũ cognitione multùm breui ſis profecturus.

Appendices habentes.

Olor, anſer, anas, gallinaceus, perdix, ciconia, aſcalaphus, tarda, noctua, paſſer.

Ingluuies habentes.

Gallinaceus, palumbes, perdix & columbus.

Gulas totas amplas habentes.

Anſer, anas, gauia, cataracta, & tarda.

Gregales aues.

Olor, anſer minor, grus, & platea.

Frugibus uictitantes.

Palumbes, columbus, turtur, & uinago.

[p. 148]

Lacus frequentantes.

Ardeola, albardeola, ciconia, gauia cineria, iunco, cinclus, trynga, calidris, & alcedones.

Mare frequentantes.

Alcedo, carulus, gauia alba, fulica, mergus, rupex & cataracta.

COMMON PLACES, REFERRING TO BIRDS, FROM ARISTOTLE.

IF it should happen that you read Pliny, Aristotle, Aristophanes, or any other suitable writer, it is fitting to have ready for use such Common Places of this kind as I will add for the sake of example only, written in a note-book, that you may with certainty refer to all those birds, of which you read in their pages. And if you do this, I doubt not that in a short time you will make great progress in the knowledge of birds.

Those having appendices[1].

Olor, anser, anas, gallinaceus, perdix, ciconia, ascalaphus, tarda, noctua, passer.

Those having craws.

Gallinaceus, palumbes, perdix, and columbus.

Those having wide gullets.

Anser, anas, gavia, cataracta, and tarda.

Gregarious birds.

Olor, the smaller anser, grus, and platea.

Those living on crops.

Palumbes, columbus, turtur, and vinago.

Those frequenting lakes.

Ardeola, albardeola, ciconia, the grey gavia, junco, cinclus, trynga, calidris, and alcedones.

Those frequenting the sea.

Alcedo, carulus[2], the white gavia, fulica, mergus, rupex[3], and cataracta.

[1] That is *cæca* or blind-guts.
[2] κύανος. [3] Perhaps χαραδριός.

Amnes & lacus frequentantes.

Olor, anas, phalaris, urinatrix, boſca, coruus palmipes, uterque anſer, uulpanſer, capella, penelops, aquila marina.

Spinas appetentes.

Spinus, carduelis, & auriuittis.

Culicibus uictitantes.

Pici Martij, galgulus, culicilega.

Vermibus aut ex toto aut magna ex parte uictitantes.

Fringilla, paſſer, rubetra, luteola, & pari omnes, ficedula, atricapilla, rubicilla, rubecula, ſyluia, curuca, aſylus, florus, montifringilla, regulus & frugilega.

Plures locos cuique licebit huius modi excogitare.

Those frequenting rivers and lakes.

Olor, anas, phalaris, urinatrix, bosca, the web-footed corvus, either kind of anser, vulpanser, capella, penelops, the sea aquila.

Those feeding on thistles.

Spinus, carduelis, and aurivittis.

Those feeding on insects.

Pici martii, galgulus, culicilega.

Those feeding on worms, either wholly, or for the most part.

Fringilla, passer, rubetra, luteola, and all the pari, ficedula, atricapilla, rubicilla, rubecula, sylvia, curuca, asylus, florus, montifringilla, regulus, and frugilega.

Any one may devise more Places of this sort.

[p. 149]

PERORATIO AD LECTOREM.

NON deerunt forſan, qui mihi hoc uicio uerſuri ſunt, quòd libellus iſte meus coniecturarum multò plus quàm certarum adſertionum in ſe contineat: quibus reſpondeo, in re ardua, & nondum ſatis explorata mihi conſultius & modeſtius uideri, hęſitanter & modeſtè coniectando ueſtigare, & ita inquirere, quàm temere & impudenter de rebus incompertis pronunciare. Quòd autem de moribus & medicinis auium nihil hic ſcripſerim, in cauſa fuerunt, temporis anni infœlicitas, & anguſtia (breuiori enim ſpacio quàm duobus menſibus totus liber [p. 150] conſcriptus eſt) & pecuniæ copia minor, quàm quæ huiuſmodi negocio abſoluendo ſufficeret. Nam quis ſine magna pecuniæ ui in longinquas regiones proficiſci poteſt, peregrinarum auium formas & mores contemplaturus, & illic diu ea de cauſa manſurus? Quis familia aut uocatione ſua, aut alijs negocijs foras prodire prohibitus, ſine maximis impenſis omnia auium genera ab aucupibus ad ſe ex uarijs mũdi plagis allata, curare poterit? et iã allata, quò mores ad plenum perueſtiget, in uiuarijs & caueis ſine maximis ſumptibus alere quis sufficiet? Hoc

PERORATION TO THE READER.

THERE perhaps will not be wanting those who will attribute this to me as a fault, that this little book of mine contains within it many more conjectures than sure statements: to whom I reply, that it seemed to me much more prudent and becoming on a subject that is difficult and not yet sufficiently explored to tread doubtingly and modestly by conjecture, and so to enquire, than to pronounce rashly and immodestly on things undetermined. Moreover that I have written nothing here of the habits and medicinal nature of birds, I have for reasons the unsuitability of the time of year and its brief span (for the whole book was written in a space of less than two months) and a supply of money too slender to suffice for the perfection of a work of that kind. For who without great command of money can set off for distant regions, to observe the forms and habits of foreign birds, and there to stay a long time for that purpose? Who, hindered by his household or his calling or other business from going abroad could without vast expense give heed to all the kinds of birds brought to him by fowlers from the various quarters of the world? and when brought, who would be capable without vast expense of keeping them in vivaria and cages, that he might investigate their habits to the full?

Alexander ille omnium ethnicorum regum potentia, bellica gloria, & literarum ſtudio maximus & nobiliſſimus ſecũ animo perpendens, [p. 151] Ariſtotelem iam de animalibus ſcripturũ, quem priuatis ſumptibus negocium illud abſoluere non potuiſſe cognouerat, ad conducendos aucupes, & uenatores, & ad alenda in uiuarijs animalia iam capta, 480 milibus coronatorum donauit, & ita inſtruxit. Talis ſi hodie alicubi Alexander exiſteret, non dubitarem quin nouus nobis Ariſtoteles alicunde renaſceretur, qui prioris illius Ariſtotelis animalia omnia, paucis exceptis, nõ ſolùm nobis cum moribus medicis ſuis facultatibus, & huius tẽporis nominibus exhiberet, ſed & multa plura animalia quã prior, & humano generi non minùs utilia nos doceret. Quòd igitur Ariſtoteles de tam multis animantibus [p. 152] tã fœliciter ſcripſerit, Alexãdri potiùs liberalitati quàm Ariſtotel. diligentię tribuendum erat: quamuis et ea ſummopere laudãda erat. Nam ſi Alexandri munificentia Ariſtoteli animalia illa ſpectanda nõ exhibuiſſet, hiſtoriam animalium nobis tam abſolutam nunquam ædidiſſet. Mirari igitur deſinant ſcriptorum huius temporis ocioſi cẽſores, frigidius, indoctius, & minori cum digẽtia ſtirpium, auium, piſcium & quadrupedum hiſtorias hac ætate cõſcribi, quàm apud ſeculum priùs tractabantur. Quum hodie quicquid pręclari in lucem emittitur, priuatis tenuioris fortunæ ſtudioſorũ uirorum ſumptibus & typographorum impenſis edatur. Sed ad te iam redeo, candidiſſime lector, quem etiam atq; etiam obteſtor, ut ſi qua tibi ſeſe peregrina facie offerat [p. 153]

The well-known Alexander, the greatest and most renowned of all kings of the nations in power, warlike glory, and zeal for learning, weighing this in his mind, presented Aristotle with 480,000 crowns, when he was about to write on animals, since he knew that the philosopher could not carry out that task with his private means, for the purpose of hiring fowlers and hunters, and for keeping in vivaria the animals which had been already caught, and provided to that end. If such an Alexander existed anywhere to-day, I should not doubt that a new Aristotle would be born again for us from somewhere, who not only would display to us all the animals, with few exceptions, of that former Aristotle, with their habits, their medicinal properties, and their latter-day names, but would inform us of many more animals than the former, and those not less useful to the human race. That Aristotle therefore wrote so happily about so many living creatures is to be put down rather to the liberality of Alexander than to the diligence of Aristotle: though that too must be praised without stint. For if the bounty of Alexander had not supplied to Aristotle those animals to be examined, he never would have published so complete a History of Animals for us. Therefore let the ease-loving critics of the present day cease to wonder that the histories of plants, birds, fishes, and quadrupeds are written in this age with less spirit, less learning, and less diligence than that with which they were treated in a former age. To-day whatever of value is brought to light is published at the private expense of very zealous men of slender fortunes and printers. But now I return to you, most ingenuous reader, and beseech you once and again that, if any bird of foreign aspect meet

auis, paucis mihi illam, addito nomine gentis tuæ, depingere non dedigneris, & mihi & omnium bonarũ literarum ſtudioſis ſcies te magnopere gratificaturum, nam nec facti in ſecunda huius libri editione ero immemor, nec quicquã, quod ad hoc diſciplinæ genus pertinet, mihi exploratum, te celabo. Vale. Coloniæ Calend. Martijs.

your eye, you will not disdain to depict it for me in a few words, with the addition of the name of your family, and you will know that you will greatly gratify both me and those who are zealous of all good learning, for I shall be neither unmindful of the act in a second edition of this book, nor conceal from you anything, which pertains to this kind of teaching, and is found out by me.

Farewell. At Cullen. 1st March.

[p. 154] ΑΛΒΕΡΤΟΣ ̔Ο ΓΕΛΡΙΕΤΣ

τῷ τῆς βίβλου σπουδαίῳ ἀναγνωστῃ
ἓυ πράττειν.

Ποίκιλα ἐι ἐθέλεις πτην' ἐιδέναι ὦ φιλόμουσε
Τοῦτο ποίημα βραχὺ τάχ' ἀνάγνωσον ἄγε.
Ουδεὶς ἐστ' ἀκριβῶς περὶ τούτων γράψεν ἰατρῶν,
Ωστε βλέπεις τῇ δ' ἐν πάντα γραφέντα βίβλῳ.
Τοῦ ὀῦν σπουδαίου Τουρνήρου τὸν πόνον οὗτον
Υπέλαβεν: τούτου καὶ ἀπόλαυε καλῶς.

Ἄλλο.

Φωνὰς μανθανέμεν πτηνῶν χιλίας ἀναγνωστά
Βούλεις Τουρνήρου τὴν βίβλον ὠνέεο.
Τοῖα γὰρ οὐκ ἰατρῶν μηδεὶς πρὶν ἐγράψατο πάντων,
Εν βιβλίῳ τούτῳ, ποῖα δοθέντα βλέπεις.
Ει ἄρ' ἁβρῶς τε σαφῶς, τῆς γραφθείσης ἀπολαύειν
Βίβλου ἀνδάνει, κίνεε δεῦρο πόδον.

Aliud eiuſdem ad candidum lectorem.

Accipe quæ docti medici tibi cura parauit
[p. 155] *Turneri, notas quiſquis auebis aues.*
Has tibi tam uarijs manus ingenioſa figuris
Expreſſit, noſſe ut quamlibet inde queas.
Ne quoque non poſſes has pernouiſſe, Latinas,
Anglas, Teutonicas, Argolicasq; facit.

Liber ad lectorem.

Quiſquis aues uarias de nomine deq; figuris
Nôſſe cupis lector, me lege, doctus eris.
Nec dabo, crede mihi, tibi munera parua laboris:
Nam uolucrum res eſt maxima ſcire genus.

Crebrò Grãmatici hîc hærẽt, ſtãt κωφὰ πρόσωπα,
Nec facit officio ſtultula turba ſatis.
Hîc ipſos medicos errare miſerrima res eſt,
Quos decet hæc animis nota tenere ſuis.
Seu Maro ſit pueris, ſeu Naſo poëta legendus,
Seu fuerint quæuis ſcripta legenda tibi:
Diſpeream, ſi non multò tibi maximus error
Occurret paſſim, ni bene nôris aues.
Autorum nimiam placet haud poſuiſſe cateruam.
Vt tibi, quæ teneo, noſtra probare queam.
Plinius hîc ille eſt, & Ariſtoteles, reliquiq́;,
Quotquot de uolucrum nos ratione docent.
Hoc ſcio, Turneri miraberis ipſe laborem, [p. 156]
Doctrinam, ſummam cum pietate fidem.

Αδηλον.

En tibi, quos docti dedit hîc pia cura labores
Turneri medici, candide lector habe.
Inuenies nimium quæ te didiciſſe iuuabit,
Hinc uenient fructus in tua uota boni.

FINIS.

APPENDIX.

EXCERPTA EX OPERE IOANNIS CAII BRITANNI DE RARIORUM ANIMALIUM ATQUE STIRPIUM HISTORIA, fol. 17—23.

De Haliaeto.

De Auibus.

HALIAETOS, id genus aquilæ eſt, quod ex mari lacubusq; prędam quęrit, vnde nomen inuenit. Is magnitudine Milui eſt, capite albis & fuſcis distincto [f. 17 b] lineis, vt melino: roſtro aquilino: oculis in medio nigris, in ambitu aureis: lingua ferè humana, niſi quòd ad radicem vtrinq; habet appendicem: colore per ſumma aſturis, per ima albo: gutture maculis notato ruffis vt & ventre, pectore medio pure candido: crure craſſo & ſquamoſo: pede vncungui & cęruleo: digitis quatuor, per ſuperna ad dimidiam longitudinem etiam ſquamoſis, ad reliquam inciſis, per inferna aſperis & aculeatis tenacitatis cauſa: & his tam validis, vt flexos vix vlla vi extendas. Prędator is eſt piſcium, diſcuſsis decidentis corporis impetu aquis, ex eisq́ue viuit. Et quamuis ex piſce viuat, fidipes tamen eſt vtroque pede, non altero palmipes, vt vulgus putat. Giraldus Cambrenſis libro de Topographia Hiberniæ, vbi de auibus biformibus agit, hunc Aurifriſiū vocat: & altero pede aperto & vnguibus armato eſſe, altero clauſo cum vulgo ſcribit. Supra magnitudinem corporis alæ longitudo eſt, quæ ad pedes Romanos duos & digitos vndecim extenditur. Inoleuit opinio iſtic apud noſtrum vulgus in Britannia, eam ineſſe vim naturalem huic aui, vt quem conſpexerit piſcem, eum ſe quàm mox reſupinare & conuertere, atque ad ſum-

APPENDIX.

EXTRACTS FROM THE WORK OF JOHN CAIUS 'DE RARIORUM ANIMALIUM ATQUE STIRPIUM HISTORIA' (1570).

OF THE SEA EAGLE.

THE Haliaetos is that kind of Eagle, which seeks its prey from the sea and lakes, whence it takes its name. It is of the size of a Kite, having the head marked with white and dusky lines, as in a badger; an Eagle's beak; eyes black in the middle, golden in the outer circle; a tongue almost like that of man, except that at the root it has an appendage on either side; the colour above that of a Goshawk, white below; the throat marked with rufous spots, as is the belly; the middle of the breast pure white; the legs thick and scaly; the foot with curved claws and blue; four toes scaly above for quite half of their length, fissured for the rest, rough on the lower part and sharp for a firm hold; and these so strong that you can scarcely straighten them by any force when bent. This bird is a preyer upon fishes, the water being cleft by the shock of its body as it plunges, and on them it lives. And though it lives on fishes, yet it is cloven on each foot, not webbed on one as the vulgar think. Giraldus Cambrensis in his book on the Topography of Ireland, when he treats of unequally formed birds, calls this the Aurifrisius[1]; and writes in common with the vulgar that it has one foot free-toed and armed with claws, the other closed (with webs). The length of the wing surpasses that of the body, for it extends to two Roman feet and eleven inches. In this affair an opinion has grown up among our common people in Britain, that such a natural power exists in this bird that any fish which it sees turns upwards on its back as soon as possible and rises to the surface

[1] Aurifrisius must be the Latinized form of the old French name 'Orfraie'—which, like Osprey, is a corruption of *Ossifraga*.

mam aquam aſcendere, in eaq́ue fluitare vt ſopitum, quo facilior prȩda ſit volanti. Ideoq́, eius adeps studioſius aſſeruatur à noſtris piſcatoribus, quòd eandem vim habere creditur. Frequentes ſunt apud nos in maritimis locis & Vecti inſula. Noſtri an Oſprey vocant. Moribus placidus eſt & tractabilis, & famis [f. 18] patientiſsimus. Vixit enim ſeptem dies apud me ſine cibo, & in alta quiete: niſi ſi hoc non mos fecit ſed fames, quæ omnia domat. Carnem oblatam recuſauit: piſcem non obtuli, quòd eum ex hoc viuere didici. Caro illi nigra eſt.

De anſere Brendino.

Anſer Brendinus, auis eſt marina, palmipes, figura anſeris, ſed magnitudine paulo infra, capite albo exiguo & curto, ſed roſtro nigro, à quo linea nigra ad oculum vtrumque ducitur, collo fuſco, à pectore ad caudam ex dimidio corpore inferiori albo, coxendice murino (vt eſt Columbæ vulgaris color aut gruis) ex ſuperiori, ad collum fuſco, vt & ad caudam medio inter vtrumq́, murino: alis item murinis, cum cauda longitudine paribus, ſed pennis ad extremum obfuſcatis. Cauda nigra eſt ex albo enata, pede nigro & palmato. Gregalis auis eſt & garrula. Ex piſce viuit, frequens apud nos per littora in Britannia. Vulgus Britannorum quod ad littus habitat à coloris varietate a Brendgoſe nominat. Ornithopolæ Londinenſes Bernaclum vocant, cum dicendum putem Berndclacum ſeu Brendclacum, quòd antiqui Britanni atque item Scoti, anſeres marinos, paluſtres & lacuſtres omnes Clakes dicebant, cum tamen hodie corruptè dicimus Fenlakes & Fenlagges, cum dicendum fuit Fenclakes. Varium item colorem, Brend, ſeu per metatheſim Bernd ijdem appellabant. Vnde [f. 18 b] bernded ſeu brended id animal dicitur, quod in colore murino variegatum eſt albo, vt eſt hic anſer. Non eſt

of the water, and therein floats as if stunned, in order that it may more readily become a prey to the winged creature. And therefore is its fat preserved with greater keenness by our fishermen because it is believed to have the same virtue. They are abundant with us on sea-coasts and in the Isle of Wight. Our people call it an Osprey. In its habits it is quiet and amenable, and most patient of hunger. For it has lived with me for seven days without food, in deep repose; unless indeed it was not habit that caused this, but hunger, which tames all things. It refused flesh when offered to it; I did not offer fish, because I understood that it lived on it. Its flesh is dark.

OF THE BRENT GOOSE.

The Brent Goose is a web-footed sea-bird, of the appearance of a Goose, but a little less in size, with the head small and short and white, but the beak black, from which a black line reaches to either eye; the neck dusky, the lower half of the body from the breast to the tail white; the flank mouse-coloured above (like the colour of a common Pigeon or Crane), dusky towards the neck, and also the intermediate parts towards the tail mouse-coloured; the wings also mouse-coloured, equal in length to the tail, but with the feathers darkened at the tip. The tail is black from a white base, the foot black and webbed. It is a gregarious and noisy bird. It lives on fish, and is common with us in Britain about the coasts. The common people of the Britons who live on the coast, name it a Brendgose from the varied nature of its colouring. The London bird-dealers call it Bernacle, yet I should think that it ought to be named Berndclac or Brendclac, because the Britons of old, as also the Scots, called all the Geese whether of the sea, marshes or lakes "Clakes," though at the present day we corruptly say Fenlakes and Fenlagges[1], when we should say Fenclakes. The same people, moreover, call anything variegated Brend or by metathesis Bernd. Whence an animal is said to be bernded or brended which is variegated with white upon a mouse-coloured ground[2], like this Goose. It is not, there-

[1] According to this notion 'Grey-Lag' would be for 'Grey-Clak.'

[2] Possibly, then, 'Bergander' is for 'Bernd-gander.'

Anſer Baſſanus. ergo Scotorum Baſſanus anſer, qui in Baſſe Scotorum Inſula nidum ponit atq; oua, à qua nomen habet. In hanc inſulã rupem exiſtentem, nec in ſummo quantam Miluus oberret (vt Poëta dixit) ſed exiguam, venturi ſtato anni tempore anſeres, quo prius ſpeculatu, qua obſeruatione prę̄miſsis nuntijs vtantur quàm ingrediuntur: quo anni tempore hoc faciant, qua ſolitudine inſulæ, concludentibus ſe incolis ad aliquot dies, donec ſe firmauerint anſeres, ne abigant, quanta multitudine atq; denſitate inuolent, ſic vt in ſerenitate ſolem adimant, quot piſces afferant, quot oua pariant, & quantos fructus in annos ex eis anſerumq; plumis atq; oleo percipiant inſulani (nam Pupinorum pinguedinem habent atq; guſtũ) longum eſſet recenſere.

De Anate Indica.

Eſt apud nos ex India anas, eadem planè corporis figura, eodem roſtro & pede quo vulgaris, ſed ex dimidio maior ea & grauior. Caput illi rubeſcit vt ſanguis, & bona pars coniuncti colli à poſteriore parte. Id totum calloſa caro eſt, & inciſuris diſtincta: quaq; ad nares finit, carunculam demittit à reliqua carne figura ſeparatam, qualis cygnis eſt, roſtro coniunctam. Nudum plumis caput eſt, & ea quoque colli pars quæ rubeſcit, niſi quòd in ſummo capite criſta eſt plumea atque candida, per totam capitis longitudinem pro- [f. 19] tenſa: quam, cum excandeſcit, erigit. Sub oculis ad roſtri initium per inferna, inordinatæ maculæ nigræ carni ſunt inductæ: & vna atq; altera à ſummo oculo ad ſuperna eleuatæ. Oculus flaueſcit, ſeparatus à reliquo capite circulo nigro. Sub extremo oculo in auerſum macula eſt ſingularis, ſeparata à cę̄teris. Roſtrum totum eſt cœruleum, niſi quòd in extremo macula nigreſcit vna. Pluma illi per totum colli proceſſum reliquum, alba. Qua corpori collum iũgitur, circulus eſt plumeus niger, rara pluma alba, maculoſus & inę̄qualis, per ima anguſtior, per ſumma latior. Poſt eum per totum imum ventrem pluma

fore, the Bass Goose of the Scots, which has its nest and eggs on the Bass, a Scottish Isle, and thence takes its name. Now when at a certain season of the year the Geese are about to return to this precipitous island rock—not so big on the top as a Kite could hover over (as the Poet has said), but very small—it would be too long to recount what spying, what circumspection (scouts having been sent ahead) they use before they alight: at what time of year they do this, the solitary state of the isle, when the inhabitants shut themselves up for several days, until the Geese have settled down, lest they should drive them off, in what numbers and in what a throng they fly to it, so that in clear weather they obscure the sun, how many fishes they bring home, how many eggs they lay, and what profit the dwellers in the isle make annually from the feathers and the oil of these Geese (for they possess the fatness and the taste of Pupins).

OF THE INDIAN DUCK.

There is among us, a Duck from India, with exactly the same form of body, the same beak and foot as the common bird, but bigger by half and heavier. Its head is red as blood, as is a good part of the adjoining neck behind. The whole of this is callous flesh and marked with fissures: and where it ends at the nostrils it makes a caruncle like that in Swans, separated in form from the rest of the flesh, which joins the beak. The head and red part of the neck are devoid of feathers, save that on the top of the head is a white feathery crest, extending over the whole length of the head; and this the bird erects, when it is excited. Under the eyes to the beginning of the beak at its lower part irregular black spots are arranged on the flesh: and one or two reach upward from the top of the eye to the parts above. The eye is yellowish, being separated by a black ring from the rest of the head. Close behind the eye is a solitary mark, apart from the rest. The whole beak is blue save that at the tip one spot shews black. The plumage over all the rest of the neck is white. Where the neck joins the body, there is a ring of black feathers spotted and irregular—with an occasional white one—narrower below, broader above. Behind this the plumage is white over the whole of the belly below:

alba eſt: per ſummum corpus, fuſca, ſed ab circulo illo nigro pluma alba in ſummo diuiſa. Extremæ alæ atque cauda cum ſplendore vireſcunt, vt Cantharides. Tibiarum cutis fuſca eſt, inciſuris leuibus per circuitus ducta. Membrana per interualla digitorum pedis palleſcit magis, vna atque altera reſperſa macula fuſca, incerta lege diſpoſita, niſi in interuallo ſiniſtri pedis, vbi ſex per digiti extremi longitudinẽ diſponuntur. Tardo gradu incedit propter corporis grauitatem. Vox illi non qualis cęteris anatibus, ſed rauca, qualis faucibus humanis catarrho obſeſsis. Mas maior eſt quàm fœmina. Ea ſimilis mari eſt, niſi quòd non ita variegato corporis colore eſt. Viuit ex cœnoſis aquis, & alijs quibus cętera vulgaris anas gaudet.

[f. 19 b] *De Anate Turcica ſiue Indica altera.*

Anati quidem ſimilis eſt quæ Turcica ſiue Indica dicitur, ſed quantitate & magnitudine corporis anſerem ferè diceres. Tota eſt candida, niſi quòd roſtrum, tibiæ, atque pedes rubent, genæq; item calloſa carne, & roſtri tuber ſupra nares. Caro illi dulcis eſt, & vox ſibilus. Sunt eius generis quędam, colore albo & nigro variegatæ. In aqua viuit, lociſq; gaudet cœnoſis vt cęteræ anates.

De Pica marina.

Pica marina (vt noſtrum vulgus nominat) paulo maior eſt terreſtri, coloris quidem varietate in corpore ſimilis, ſed colore pedis, digitorum numero, inciſuris, cauda atque roſtro admodum diſsimilis. Nam pes rubet, & digito poſteriori deſtituitur, nec inciſuras habet is & tibia, ſed impreſsiones quaſdã, piſcium ſquamis quàm ſimillimas. Fiſſus eſt, ſed ita in digitis vtrinq; luxuriat cutis, quemadmodũ in fulicis penè, vt ad natandũ quoque pes factus videatur. Auis eſt Amphibios. Caudam curtam habet, roſtrum longum atq; tenue, perpendiculariter latum, non teres, colore

and dusky over the upper surface of the body, but the white feathering above is divided by the black ring. The ends of the wings and the tail have a greenish gloss as in Cantharides. The skin of the legs is dusky, marked all round with slight fissures. The web in the spaces between the toes is paler, marked now and again with a dusky spot, arranged in no precise plan, except in those of the left foot, where six are distributed over the length of the outer toe. The bird moves with slow step on account of the weight of its body. Its voice is not like that of other Ducks, but hoarse; such (as comes) from the human throat when attacked by a catarrh. The male is larger than the female. She is like the male, save that she has not so varied a colouring of body. The bird gets its living from muddy waters, and those others wherein the other common Duck delights.

Of the Turkish or Second Indian Duck.

That which is called the Turkish or Indian is like a Duck, but (judging) from the bulk and bigness of its body you would almost call it a Goose. It is entirely white, except that the beak, legs and feet are red, while the cheeks also have callous skin, and there is a protuberance on the beak above the nostrils. Its flesh is sweet, and its voice whistling. There are some of this kind variegated with black and white. It lives upon the water, and delights in muddy places, as do other Ducks.

Of the Sea Pie.

The Sea Pie (as our common people call it) is a little larger than the land Pie, yet like it in the varied colour of its body, while very dissimilar in the colour of the foot, the number of toes, their scutellations, the tail and the beak. For the foot is red and lacks a hind toe, nor has that member or the tibia scutellations, but merely marks, as like as may be to the scales of fishes. It is cloven-footed, but there is such an extent of skin on the toes on each side—almost as much as in Coots—that the foot would even seem to be formed for swimming. It is an amphibious bird. It has a short tail, a beak long and thin, vertically broad, not smooth, in colour

in ſummo ad caput rubrum, per reliquum pallidum, nec eſt in fine acutum, ſed obtuſum id. In menſa grata auis eſt.

De gallina Getula domeſtica.

Gallina Getula domeſtica, paulo minor noſtra eſt, [f. 20] colore in supernis obſcurè ruffa, in infernis pallida, pluma capitis incompoſita & erecta, criſta ſerrata, humili, ſimplici, carnea: gutturoſa magis perpetuò, quàm noſtræ cum glociunt: tibijs pedibusq́ue plumoſis, maximè per exteriora & poſteriora vt columbis, ne per interiores plumas greſſus impedirentur: cętera vt vulgaris.

De Meleagride.

Meleagris pulcherrima auis eſt, magnitudine corporis, figura, roſtro, & pede Phaſiano ſimilis: vertice corneo, in apicem corneum à poſteriori parte pręcipitem, in anteriori leniter accliuem eleuato, armatus. Eum natura voluiſſe videtur inferiori capitis parti tribus veluti lacinijs ſe promittentibus committere atque deligare: inter oculum & aurem vtrinque vna, & in fronte media item vna: omnibus eiuſdem cum vertice coloris: ita vt inſideat capiti eo modo, quo ducalis pileus illuſtriſsimo duci Veneto, ſi quod iam aduerſum eſt, auerſum fieret. Rugoſus is eſt: inferius, per circuitum: qua ſe attolit, in directum. In ſummo collo ad occipitium naſcuntur erecti quidam atque nigri pili (non plumæ) in contrarium verſi. Oculi toti nigri, æquè & in orbem palpebræ atq; cilia, ſi maculam in ſumma & poſteriori parte ſupercilij vtriuſque demas. Imum caput per longitudinem vtrinque caro quędam calloſa colore ſanguineo occupat, quæ ne propendeat veluti Galli gallinacei palea[1], vt [f. 20 b] replicaretur natura voluit, & auerſo ductu in duos proceſſus acutos à capite liberos finiret. Ex hac carne attollunt ſe vtrinque carunculæ, quibus nares in ambitu veſtiuntur, & caput in anteriori parte à cętero roſtro pallido ſeparatur. Harũ ad roſtrum margines inferiores, replicantur etiam leuiter ſub vtroque nare.

[1] Possibly 'palea' is a misprint for 'galea,' comb.

red on the top near the head, pale on the remainder, nor is it sharp at the end, but blunt. It is an acceptable bird for the table.

Of the domestic Getulian Hen.

The domestic Getulian Hen, is a little smaller than our own, in colour dull rufous above, pale below, with an erect crest of disintegrated feathers on the head, a serrated comb, low, simple and fleshy: more continuously noisy than ours are, when they cluck: with the legs and feet feathered, for the most part outwardly and behind as in Doves, that the progress should not be hindered by feathers on the inside, otherwise it is like the common kind.

Of the Meleagris.

The Meleagris is a very beautiful bird, like to a Pheasant in bigness of body, form, beak and foot: provided with a horny poll rising to an abrupt horny peak at the back, which slopes down gently in front. Nature seems to have designed to join and bind this to the lower part of the head by three hanging lappets as it were; one on each side between the eye and the ear, and also one on the middle of the forehead, all of the same colour as the poll, so that it sits on the head in the same way that the ducal cap does on that of the most noble Duke of Venice, if that part which is usually in front be turned behind. It is wrinkled round about below, but vertically where it rises above. From the top of the neck to the occiput spring certain erect black bristles (not feathers), turned backwards. The eyes are wholly black, and equally so are the eyelids and eyelashes around them, if you except a mark on the top and back of each eyebrow. A kind of callous flesh of a blood-red colour covers the lower part of the head along its length; nature has designed that it should be folded, and should not hang forward like the wattle of the Fowl, and being led backward end in two acute processes free from the head. From this flesh rise on either side caruncles, by which the nostrils are clothed round about, and by which the head is divided in front from the rest of the pale-coloured beak. The lower edges of these by the beak are also folded slightly under each nostril. What intervenes

Quod inter verticem & carnem eſt à dextra & ſiniſtra parte, album deplume eſt, leui cœruleo mixtum. Color verticis atque apicis, idem prorſus eſt cũ colore dactyli. Tibiæ nigræ ſunt, & in anteriori parte, ſquamoſa inciſura duplici notatæ: in poſteriori nulla, ſed leues, & veluti punctis quibuſdam ſui coloris reſperſæ. Color illi ſub faucibus exquiſitè eſt purpureus: in collo obſcurè purpureus: in cętero corpore per ſumma contuenti, qualis conſurgit ſi album & nigrum pollinem vtcunque tenuiter tritum, colori fuſco rarius aſpergas, nec tamen commiſceas. Tali colori maculæ albæ ouales aut rotundæ per totum corpus ineſſe viſuntur, per ſumma minores, per ima maiores, comprehenſæ interuallis linearum (vt apparet in plumarum compoſitione naturali) quæ ſe mutuo interſecant obliquo hinc inde ductu, per ſumma tantum corporis, non item per ima. Id non ex toto corpore ſolum deprehendes, ſed ex ſingulis auulſis plumis. Superiores enim, obliquis lineis ſe mutuo interſecantibus, aut, ſi mauis, orbiculis quibuſdam ex albo & nigro (vt dixi) polline confectis, & per extremitates coniunctis, vt in fauis aut rhetibus, maculas ouales aut [f. 21] rotundas albas in ſpacijs fuſcis comprehendunt: inferiores non item. Vtręq; tamen ſimili lege poſitæ ſunt. Nam in alijs plumis, ordine ita iunctæ ſunt, vt ferè triangulos acutos faciant: in alijs, vt oualem figurã repręſentent. Huius generis ordines tres aut quatuor in ſingulis ſuis plumis ſunt, ita vt minores in maiorũ complexu reponantur. In extremis alis & in cauda, rectis lineis ęquidiſtantibus procedunt per longitudinem maculæ. Inter gallum & gallinam vix diſcernes, tanta eſt ſimilitudo, niſi quòd gallinæ caput totum nigrum eſt. Vox illi eſt diuiſus ſibilus, non ſonorior, non maior voce coturnicis, ſed ſimilior voci perdicis, niſi quòd ſubmiſsior ea eſt, nec ita clara. Curſu velox eſt.

De Morinello.

Morinellus, auis nobis cum Morinis communis, ſtulta admodum eſt, ſed in cibis delicata, eoq́ue apud

between the poll and the flesh on the right and left is white and unfeathered, mixed with light blue. The colour of the poll and peak of the head is exactly the same as that of the toes. The legs are black, marked in front with a double scaly fissure, on the back with none, but smooth, and sprinkled as it were with some spots, of a peculiar colour. The colour below the jaws is exquisite purple; on the neck dark purple; on the rest of the body, if you look from above, it stands out as if you sprinkled black and white flour, ground very fine with dusky colour at intervals, and yet did not mix them up. On such a ground oval or round white spots seem to be imposed throughout the whole body, smaller above and larger below, arranged in lines at intervals (as appears in the natural structure of feathers) which cut one another here and there in reciprocal oblique arrangement, only on the top of the body, however, and not below. This you will observe not only from the body as a whole, but from individual feathers if plucked out. For the upper feathers, with their oblique lines cutting one another reciprocally, or, if you will it, with certain rounds composed of black and white flour (as I have said), and meeting towards the tips, as in honey-combs or nets, enclose oval or round white spots within dusky spaces: but the lower do not. Both, however, are arranged by a similar law, for on some feathers they are so joined in rows as to almost make acute triangles; in others so as to present an oval form. Of this kind there are three or four rows on each feather, so that the smaller are contained in the compass of the larger. At the tips of the wings and on the tail, spots lie along the length in equidistant straight lines. You can hardly distinguish between the cock and the hen, so great is the likeness, save that the head of the hen is wholly black. The voice is a shrill double cry, not more sonorous or louder than that of the Quail, but like that of the Partridge, except that the latter is lower, and not so clear. In running it is swift.

Of the Morinellus.

The Morinellus, a bird common to us and the Morini[1], is very foolish, but delicate to eat, and on that account is a

[1] The people of a district in Northern France.

nos in ſummis delitijs atque pretio eſt. Imitatrix auis eſt. Ideo, vt Scops & Otus ſaltandi imitatione, ita hęc noctu ad lumen candelæ pro capientis geſtu capitur. Nam ſi is expandit brachium, extendit & illa alam: ſi is tibiam, & illa itidem. Breuiter quicquid gerit auceps, idem facit & ales. Ita humanis geſtibus intenta auis, ab aucipe decipitur, & rhete obuelatur. Auis parua eſt, magnitudine Sturni, tribus tantum digitis anterioribus, poſteriori nullo, vertice nigro, genis candidis, coturnicis ferè colore, ſi cinericei [f. 21 b] parum admiſceas, potiſsimum circa collum. Morinellum voco duplici de cauſa, & quòd auis eſt apud Morinos frequẽtiſsima, & quòd auis ſtolida eſt, quæ ſtultitia grȩcis μωρότης dicitur. Eam ob rem noſtri etiam Doterellũ vocant, quaſi ſtultitia delirantem dicant.

De Puphino ſiue Pupino.

Eſt auis quȩdam marina noſtras, parui anatis (quàm βόσκαδα Grȩci vocant) magnitudine & figura corporis, pedibus palmatis & rubeſcentibus, ad poſteriora magis poſitis quàm cȩteris palmipedibus exceptis pygoſcelibus: roſtro tenuiore magis latitudine ſe demittente, quàm longiore proceſſu ſe extendente, quatuor inciſuris rubris à ſumma, duabus ab ima parte ſulcato, in colore pallentis ochræ. Quod inter has & caput eſt, ſubcœruleum eſt, & ea figura qua luna eſt, cum exacti dies decem ſunt à coitu. Per ſumma corporis totius nigreſcit, niſi qua oculi ſunt, qui in albo conſtituti ſunt: per ima exalbeſcit tota, niſi ſummo pectore, qua nigricat. Viuit ex mari. Hunc noſtri Puphinum dicunt, nos Pupinum à naturali voce pupin. Latitat in cauernis, vt charadrius. Eam ob rem educta è cuniculi cauea auis hȩc eſt, loco non procul à mari poſito, à venatore quodam immiſſa viuerra. In piſcis vſu apud nos eſt in ſolenni ieiunio per [f. 22] quadrageſimam: carne & guſtu, Phocæ marinæ haud diſsimilis. Gregale animal eſt, & ſua habet latitandi tempora, vt Cuculus & Hyrundo. Oua parit in terræ

very great luxury with us, and of great price. It is a mimic. And so, as the Scops and Otus are taken by an imitation of dancing, this bird is caught at night by the light of a candle according to the motion of the captor. For if he stretches out an arm, the bird lifts a wing; if he stretches out a leg, it does likewise. In short, whatever part the fowler plays, the bird does the same. So being intent on the man's actions, it is fooled by the bird-catcher and caught in his net. It is a little bird, of the size of a Starling, with only three fore-toes and no hind-toe, a black crown, white cheeks, and colour almost that of a Quail, if you were to mix with it a little ash-colour, especially round the neck. I call it Morinellus for a double reason, both because it is a bird most abundant among the Morini, and because it is a foolish bird, foolishness being by the Greeks called *μωρότης*. On this account our people also call it Doterell, as if they were to say doating with folly.

OF THE PUPHIN OR PUPIN.

There is a certain sea-bird of our country, in size and form of body like a little Duck (which the Greeks call *βόσκας*), with webbed and reddish feet, placed nearer to the hinder parts than in other web-footed kinds except the Pygosceles: with a somewhat thin beak, rather more extended in breadth vertically than stretching laterally to a very great length, furrowed by four red grooves above, and two below, pale ochre in colour. The part lying between these and the head is bluish, and of such a shape as is the moon, when ten days have elapsed from conjunction. The bird is black on the upper surface of the whole body, save where the eyes are set, which are enclosed in white: but it is wholly white below, save on the upper breast, where it is black. It gets its living from the sea. This bird our people call the Puphin, we say Pupin from its ordinary cry of "pupin." It hides in holes, as the Charadrius does. And so it is driven out from a rabbit's burrow by a ferret turned in by any hunter in a place situated not far from the sea. It is used as fish among us during the solemn fast of Lent: being in substance and taste not unlike a Seal. It is a gregarious animal, and has its proper time for lying hidden, as the Cuckoo and Swallow.

cuniculis bina magna ex parte. Alis non confidit nisi conspecto mari. Pigrum videtur animal, sed iniurię patiens. Vescitur carne lubentius quàm pisce, & cuniculi quàm alterius animalis, sed vtroque crudo: cocta & assa respuit. Cętera non attingit humana ędulia. Aestate se lauat, sed nunquam, quod obseruatione deprehendi potest, bibit: an quòd aqua marina carebat, nescio. Excrementum alui illi est quale accipitri. Cum non esset quod ederet, cibum voce naturali geminata & summissa, pupin, pupin clamitando, implorabat. Alebam domi meæ ad menses octo. Mordebat lubenter ministrantes cibum, aut attingentes, sed benignius atque innocentius. Exiguo cibo satiatur. Non enim vorax auis est vt Coruorans noster, quem tu (Gesnere charissime) coruum aquaticum & mergum recte nominas, nostri Cormorantem corruptè dicunt, nescientes ex vocis ætymo coruum vorantem appellari debere. Quod naturaliter facit, cum illi à natura vnicum tantum intestinum sine elice (vt aiunt) rectum sit concessum, propter vehementiam caloris naturalis, citissimè absumentis quæ assumuntur omnia.

Coruorans seu Mergus.

De Spermologo seu Frugilega.

Spermologus noster à cornice nigra nulla in re differt, nisi morũ innocentia, προλόβῳ seu ingluuie, [f. 22 b] qua granum legendo continet, vt ad suos referat: (est enim gutturosus) albo callo, qui in summo rostro est ad caput: & voce, quam habet gutturalem & raucam. Vnde forsan nostris, quibus nomina rerum multa Latina sunt & Gręca (vti libro nostro de symphonia vocum Britannicarum diximus) rouce dicitur, quasi rauce Anglis, raucus Latinis sit dicendus: victu quoque à coruo differt, quòd frumento, hordeo, & cętero semine (vnde nomen σπερμολόγου inuenit apud Gręcos) vescitur. Vescitur & vermibus, vbi frugis frumentique copia non est. Hinc rustici nostri dubitant

It lays for the most part two eggs in rabbit burrows in the earth. It does not trust to its wings save in sight of the sea. It seems a lazy animal, but patient of injury. It eats flesh more readily than fish, and that of a rabbit in preference to that of any other animal, but in either case raw: it throws up what is boiled or roasted. Other human victuals it does not touch. In summer it washes itself but never drinks, so far as can be ascertained by observation; whether this was because salt water was wanting, I know not. The droppings are like those of an Accipiter. When there was nothing to eat it begged for food with its ordinary cry repeated and lowered, by calling out "pupin, pupin." I kept one at my house for eight months. It bit with right good will those who supplied it with food or touched it, but in a mild and harmless way. It was satisfied with little food. For it is not a voracious bird, as our Corvorant is, which you (dearest Gesner) rightly name *Corvus aquaticus* and *Mergus*, while our people corruptly say Cormorant, not knowing from the derivation of the word that it ought to be called the Crow that devours. And this it does naturally, since it is endowed by nature with only one intestine straight and without a coil (as they say), on account of the vehemence of the natural heat, which very quickly consumes all that it swallows.

Of the Spermologus or Frugilega.

Our Spermologus differs in nothing from a black Crow, save in the harmlessness of its habits, in the προλόβος or crop, which holds the grain as it is picked up, that it may bear it to its young (for it is pouched): in the white callus, which extends from the base of the beak to the head; and in its voice, which is guttural and harsh. Whence possibly by our people, among whom there exist many Latin and Greek names for things (as we have said in our book on the Harmony of British words) it is called Rouke, as if it should be called Rauce by the English and Raucus by the Latins: also in food it differs from the Crow in that it eats wheat, barley, and other seeds (whence it got the name of σπερμολόγος among the Greeks. It also eats worms, when there is not plenty of corn or grain. Hence our country people doubt

maior ne ſit ex eis vtilitas agris dum legunt vermen, frugis & ſementis peſtem, an inutilitas hominibus dum vorant granum, hominis nutrimentum. Tanta tamen multitudo eſt, vt legibus condemnentur: innocens alioqui auis atq; vtilis. Agricolis enim teneri adhuc ex nido, in cibo ſunt. Non niſi excelſis arboribus, idq́ue ſocietate quadam numeroq́; nidificant, cohabitant, & conſidunt.

De Sacropſittaco.

Pſittacorum plura eſſe genera obſeruaui. Quidam enim puſilli, magnitudine videlicet turdi, toto quidem corpore virides ſunt, ſed caudam longam atque gracilem, & eam aut flauam, viridem, aut puniceam habent. Quidem[1] rurſum grandes ſunt admodum, cornicis magnitudine, ex toto punicei ſeu rubri, niſi ſub imo ventre, [f. 23] extremis alis, & extima cauda, quibus partibus cum cœruleo vireſcunt. His roſtrum eſt magnū, cauum, pellucidum, & aduncum, medio tantum ſui pallidum, vtroque extremo ex parte nigrum, vt & inferior maxilla tota nigra, cuius cauitatem lingua dura & nigra cęterorum Pſittacorum modo & forma occupat. Vtrinque genæ, in cute rugoſa, figura ferè triangulari obtuſa, candicant, rubris mollium pennarum ordinibus ęquidiſtanti parallelo inductis, & alicubi etiam ſine ordine. Oculus paruus, & in albo cilio cuticulari conſtitutus, pupillam habet nigram, quam circundat aureus circulus. Digitos habet quatuor ita efformatos, vt videatur natura voluiſſe omnes anteriores feciſſe, retorſiſſe tamen duos in auerſum firmandi corporis cauſa. Hos Braſilia mittit, quos propter inſignem magnitudinem Sacropſittacos nominamus. *Sacrum.* Veteres enim quod pręclarum magnumq́ue erat, ſacrum dicebant. Vt ſacrum os, ſacram anchoram, ſacrum falconem, quem hierofalconem dicunt, ſacrum piſcem, ἱερὸν μένος, ſacram famem, & ſacrum morbum. Huius generis imaginem quam à nobis accepiſti, ſubiunge. Cęteros inter hos magnitudine medios, aliæ regiones, vt inſula Hiſpania, Aegyptus, & India ferunt, ſed colore vario. Alij enim toto corpore cinereo, caudas habent fultas, breuiores & puniceas,

[1] A misprint for Quidam.

whether their utility is greater in the fields, when they pick up the vermin, destructive to crops and seeds, or their harmfulness to men, when they devour grain, the food of man. However the number of them is so great, that they are condemned by the laws; harmless and useful birds in other respects. For, when still tender from the nest, they are used as food by country men. Only in lofty trees, and that as it were, in company and in numbers do they nest, abide and roost.

OF THE SACROPSITTACUS.

I have observed that there are many kinds of Parrots. For some that are small, namely of the size of a Thrush, have the whole body quite green, but the tail, which is long and slender, either yellow, green, or crimson. Again some are very large, of the size of a Crow, altogether crimson or red, except under the lower belly, on the ends of the wings, and on the tip of the tail, on which parts they shew greenish blue. These have a large beak, hollow, shining and hooked, pale only in its middle, partly black at each end, while the lower jaw is entirely black, the cavity being filled by the tongue, which is hard and black of the style and shape of that of other Parrots. The cheeks on both sides are whitish on the wrinkled skin, in shape almost obtusely triangular, the rows of soft red feathers being set in equidistant parallel lines, though in some places also without being in rows. The eye, small and set in a white ring of skin, has a black pupil, which is surrounded by a golden circle. It has four toes fashioned in such a way that nature seems to have intended to place all of them in front, but to have turned two of them back for the sake of supporting the body. Brasil sends us these birds, which we call Sacropsittaci on account of their remarkable size. For the ancients used to call that Sacrum, which was notable and large. As the *os sacrum*, *anchora sacra*, *falco sacer*—which they call *Hierofalco*, *piscis sacer*, *ἱερὸν μένος*, *fames sacra*, and *morbus sacer*. Compare with these the picture of this kind, which you have received from us. Other regions, as the Spanish Isle, Egypt and India, produce other kinds which are midway in size, but with various colouring. For some with the whole body grey have strong tails, some-

in extremoque lunares. Alij ex toto virides ſunt. Alij cum corpore vireant, cauda longa flaueſcunt. [f. 23 b] Mores habent omnes ſimiles, & victum communem. niſi quòd Sacropſittacus, pane bera macerato, carne, & piſce etiam veſcatur.

De Coruis albis.

Anno domini. 1548. Auguſto menſe, coruos duos candidos ex eodem nido vidi & contrectaui iſtic in Cumbria noſtræ Britanniæ, apud eiuſdem prouinciæ comitem natiuos, atque ita ad aucupium factos vt accipitres. Nam & brachio falconarij quietè inſidere, & ſoluti ad eius vocem atque ſignum vel è longinquo quàm celerrimè aduolare docti erant. Hos nihil eſt infauſtum cõſequutũ, vt albas illas hirundines, de quibus Alexander Myndius apud Ælianũ. Nam qui coruũ album notat, colorem notat: vt qui vrſum album & vulpem nigram: quorum vtrumque vidimus ex Moſcouia iſtic in Britannia. Quanquam vulpem nigram non queo dicere, etſi vulgus nigram vocet: ſed potius fuſcam, aut obſcurè griſeam reliquo corpore. Nam genas atque tibias tantum fuſcas habet, clunes atque caudam.

Vrſus albus.

Vulpes nigra.

Expletis iam quæ de volatilibus ad te ſcripſi mi Geſnere, ſuo ordine piſces conſequuntur.

what short and crimson, and crescent-shaped at the tip. Others are entirely green. Others, though they are green on the body, are yellowish on the long tail. All have like habits, and the same kind of food, save that the Sacropsittacus eats bread soaked in beer, flesh, and even fish.

OF WHITE RAVENS.

In the year 1548, in the month of August, I saw two white Ravens from the same nest, and handled them at the very place in Cumberland of our Britain, bred on the property of a lord of that county, and trained for bird-catching just like hawks. For they had been taught both to sit quietly on the arm of the falconer, and when loosed to fly as quickly as possible to his call and sign even from a distance. Nothing unlucky followed them, as in the case of those white Swallows, about which Alexander Myndius wrote according to Ælian. For he who notes a white Raven notes the colour; as he does who notes a white Bear and a black Fox; both of which I have seen here in Britain from Muscovy. Yet I can hardly call a Fox black, although the common people call it black; but rather dusky or dull grey on the rest of the body. For it has only the cheeks and the legs, with the rump and the tail, dusky.

And now those things being finished which I have written to you of flying creatures, my Gesner, the fishes follow in due order.

INDEX.

Acanthis, 52, 53
Accentor modularis? xvii
Accipiter, 14–18, 57, 66, 67, 206, 207
A. nisus, xv
A. palumbarius, 14, 15, 18, 19, 56, 57
Accipitres, 14–17, 38, 66, 67, 116–118, 138, 139, 166
Actitis hypoleucus, xvii, 57
Adler, xvi, 30
Ægithus, 74, 104, 162
Æsalo, 14–16
Æsalon, 16, 17
'Αετός, 30, 31
Africanæ or *Africæ*, see *Gallinæ*
'Αιδών, 108
'Αιγιθαλός, 130, 131
"Αιγιθος, 160, 161
'Αιγοθήλας, 48, 49
'Αιγωλιος, 116
'Αισάλων, 16, 17
"Αιθηα, 110, 111
'Ακανθίς, 104, 105
'Ακανθοφάγα, 40
Alauda, xiv, 80, 81, 106, 107, 146, 147
A. arborea, xiv
A. arvensis, xiv
Alaudidæ, xiv
Albardeola, 38–40, 180, 181
Albicilla, 28, 30, 31
Alcedinidæ, xiv
Alcedo, 18–20, 22, 180, 181
A. ispida, xiv
'Αλέκτωρ, 82, 83
'Αλέκτορις, 82, 83
Älke, xv, 92, 93
'Αλκυών, 18, 19
Amsel, xviii, 114, 115
Anas, xii, xiv, 22, 23, 32, 48, 92, 93, 176, 180–183, 198, 204
A. boscas, xv, 47
A. indica, 196
A. turcica, 198
Anataria, 30–33
Anatidæ, xv
Anser, xi, xii, xv, 22–24, 26, 28, 110, 122, 180–183, 194, 196
A. bassanus, 30, 196
A. brendinus, 194
A. minor (the smaller), 180, 181
A. palustris, xi
Anthus pratensis, xvi
Apes, 98, 99
Apodes, 98–103
Apple-sheiler, 73
Aquila, xii, xv, 16, 17, 30–38, 46, 47, 56, 57, 120, 123, 128, 129, 138–141, 152, 153, 162, 163, 192
A. marina, 182
A. montana, 34
Aquila, Sea, 183
Aquila vera, 36
Aquilæ, 30, 31
Ardea, xv, 36–40, 54, 112, 124
A. alba, 38, 39
A. cinerea, xv
A. pella, 38, 39
A. stellaris, x, 38–41, 126, 127
Ardeæ, 38, 39
Ardeidæ, xv
Ardeola, 180, 181
Arlyng, xviii, 52, 53
Ärn, xvi, 30, 31
Ärn, Edel, xv, 36
Ascalaphus, 180, 181
Asio, 130, 131
A. otus, xvii
'Ασκαλώπαξ, 86, 87
Astur, 192
A. palumbarius, xv
Asylus, 182, 183
Atricapilla, 44–47, 71, 110, 111, 160, 161, 182, 183
Attagen, xvii, 40–45, 86, 92, 93, 126, 127
Attagena, 40, 41, 44, 45, 87
Atzel, xv, 142, 143
Aurifrisius, 192, 193
Aurivittis, 40, 41, 50, 51, 106–109, 182, 183
Aves Canariæ, 108

Aves Diomedeæ, 70, 71, 78
Aves tardæ, 130, 131, 168, 169 (see also *Tarda*)

Balbushard, xv, 33
Balbushardus, 32
Bald-Buzzard, 33
Βαρέα, 40
Bargander, 25
Βασιλεύς, 152, 153
Bass Goose, 31, 197
Batis, 158, 159
Βάτις, 158, 159
Bauncok, 84
Bergander, xii, xv, 24, 25, 195
Bergandrus, 24
Bergdöl, xv, 90, 91
Bernacle, 195
Bernaclus, 194
Berndclac, 195
Berndclacus, 194
Berndgander, 195
Bernicla, xi, 26, 30
B. leucopsis, xv
Bernicle (Goose), x, xv, 27, 31
Bistard, xvi, 166, 167
Bistarda, xii
Bittern, 41
Bittour, xv, 40, 41
Blackcock, ix, 43
Blak byrd, xviii, 114, 115
Blödtfinck, xvi, 160, 161
Bloudvinc, 161
Bloudvinca, 160
Βωμολόχος, 90, 91
Bonasa sylvestris, xvii
Bosca, 48, 49, 182, 183
Βόσκας, 204, 205
Botaurus, xiv
B. stellaris, xv, 41
Brachvogelchen, xviii, 158, 159
Bramlyng, xvi, 72, 73
Brant, xv, 27
Branta, xi, 26
Brech vögel, xviii, 52, 53
Brendclac, 195
Brendclacus, 194
Brend Gose, 194, 195
Brent Goose, 195
Βύας (Βρύας), 46, 47
Bubo, 46, 47
B. ignavus, xvii, 47
Bůchfink, xvi, 72, 73
Bulfinc, 161
Bulfinca, 160
Bulfinch, x
Bulfinche, xvi, 160, 161
Bunting, xvi, 135, 159
Bunting, Reed, x
Buntinga, 134, 158
Burgander, 25
Burrow Duck, 25
Bushard, xv
Busharda, 16
Bustard, xvi, 166
Buteo, xii, 14–18, 32
B. vulgaris, xv
Buttor, xv, 123
Buttora, 122
Buttour, xv, 40, 41
Buzzard, 17
Buzzard, Moor-, ix

Caddo, xv, 92, 93
Calidris, 180, 181
Canary Bird, xvi, 109
Capella, 182, 183
Caprimulgidæ, xv
Caprimulgus, xii, xiii, 48–51
C. europæus, xv
Carduelis, xii, 40, 41, 50–53, 182, 183
C. elegans, xvi
C. spinus, xvi, 51
Carulus, 180, 181
Cataracta, 70, 71, 180, 181
Cepphus, 74, 75
Certhia, 52, 53
C. familiaris, xv, 65
Certhiidæ, xv
Chaffinche (Chaffinch), xvi, 72, 73
Chalcis, 56, 57
Charadriidæ, xv
Χαραδριός, 181
Charadrius, 204, 205
C. pluvialis, xv
Χελιδών, 96, 97
Χήν, 22, 23
Chenalopex, xii, 22–25
Cheneros, 22–28, 31
Χλωρεύς, 106, 107
Χλωρίων, 106, 107, 132, 172, 173
Chloris, xii
Χλώρις, 107–109
Chogh, xv, 92, 93
Choghe, 90, 91
Choghe, Cornish, xv, 90, 91
Chough, 91
Chough, Alpine, x, 91
Chough, Cornish, x
Χρυσομῖτρις, 40, 41
Chrysomitris, 52, 53
Ciconia, x, 54, 55, 62, 63, 92, 93, 180, 181
C. alba, xv
Ciconiidæ, xv
Cicumæ, 120, 121
Cicuniæ, 120
Cinclidæ, xv
Cinclus, xiii, 54–57, 170, 171, 180, 181
C. aquaticus, xv, 23

Circus, 14-17
C. æruginosus, xv, 33
C. cyaneus, xv
Clake, 194, 195
Clanga, 30, 31
Clotburd, xviii, 52, 53
Cnipolegos, xiii
Cob, xvi, 79
Cobbi marini, 78
Coccothraustes, xiii
Cæruleo, 52, 53
Cok, xvii, 82, 83
Cok of Kynde, 84
Colius, 88, 89
Collurio, 58, 59
Columba, xv, 14, 58-60
C. œnas, xv
C. palumbus, xv
C. vulgaris, 194
Columbæ, 58, 59
Columbi, 14, 18, 42, 44, 58-61
Columbidæ, xv
Columbus, 18, 58-62, 66, 67, 180, 181
Comatibis comata, 93
C. eremita, xvi, 93
Coot, 77, 199
Copera, xiv, 80, 81
Cormorans, 206
Cormorant, ix, xvii, 110, 111, 207
Corn-Crake, 71
Cornish Choghe, xv, 90, 91
Cornix, xii, 64, 65, 90, 91, 118, 148, 208
C. aquatica, 22
C. hyberna, 64
Corvidæ, xv
Corvorans, 206
Corvorant, 207
Corvus, x, 34, 64, 65, 92, 93, 206, (*albus*) 210
Corvus (web-footed), 183
C. aquaticus, x, 92, 93, 206, 207
C. calvus, 92
C. corax, xv
C. cornix, xv, 65
C. corone, xv, 65
C. frugilegus, xv, 65
C. monedula, xv
C. nocturnus, xii, xiii
C. palmipes, 182
C. sylvaticus, x
C. vorans, 206
Corydos, 82, 83
Cotile riparia, xvi
Coturnix, 62, 63, 68, 104, 105, 126-129, 138, 139, 146, 147, 202, 204
C. communis, xviii
Coushot, xv, 60, 61
Cout, xvii, 33
Couta, 32
Crake, Corn-, 71
Crammesvogel, 172, 173
Crane, x, 55, 94, 95, 97, 195
Craspecht, xvii, 149
Craspechta, 148
Creeper, 53
Creeper, Tree-, ix
Creeper, Wall-, 53
Crepera (Creper), xv, 52
Crex, 68-71, 128, 129
C. pratensis, xvii, 71
Crow, xv, 64, 65, 119, 149, 207, 209
Crow (that devours), 207
Crow-Picus, 149
Crow, Carrion, 65
Crow, Hooded or Grey, 65
Crow, Sea, 65
Crow, Winter, xv, 68
Cryel Heron, xv, 38
Cuckoo, ix, x, 205
Cuculidæ, xv
Cuculus, 48, 49, 66-69, 204
C. canorus, xv
Cukkow, xv, 66, 67, 69
Culicilega, xiii, 64, 65, 182, 183
Curuca, 68, 69, 182, 183
Cychramus, 128, 129
Cygnus, xiii, xiv, 126, 127, 196
C. olor, xv
Cymindis, 16, 17, 56, 57
Cypselidæ, xv
Cypselus apus, xv
C. melba, xv
Cypsellus, 100, 101

Daker Hen, xvii, 70, 71, 128, 129
Daulias luscinia, xvii
Daw, 65, 77
Dendrocopus major? xvii
Dike Smouler, xvii, 136, 137
Diomede (birds of), xvii, 71, 79
Dipper, 23
Distelfinck, xvi, 40, 41
Distelvinc, 41
Distelvinca, 40
Döl, xv, 92, 93
Doterel, 205
Doterellus, 204
Douker, xvii, 176, 177
Dove, x, xv, 15, 58-61, 63, 97
Dove, Ringged, xv, 60
Dove, Venice, xv, 63
Drossel, xviii, 172, 173
Δρυοκολάπτης, 146, 147
Důcher, xvii, 110, 111, 176, 177
Duck, xv, 22, 23, 25, 33, 49, 177, 199, 205
Duck, Indian, 197
Duck, Turkish, 199
Duck, Wild, 47

Durstel, xviii, 172, 173
Duve, xv, 58, 59
Duve, Turtel, xv, 60

Eagle, 31, 33, 37, 39, 47, 193
Eagle, Sea, 193
Ebeher, xv, 54, 55
Egle, xvi, 30
Egle, Right, xv, 36
Ἐγώλιος, 116
Eissvogel, xiv, 18, 19
Elster, xv, 142, 143
Elsterspecht, xvii, 149
Elsterspechta, 148
Emberiza citrinella, xvi
E. miliaria, xvi
E. schœniclus, xvi
Endt, 22, 23
Engelchen, xvi, 108, 109
Ephemerus, 28, 29
Ephimerus, 28
Ἔποπς, 174, 175
Erithacus rubecula, xvii
Ἐρίθακος, 154, 155
Ἐριθέα, 154, 155
Erna, xii, 30
Erne, xii, xvi, 31
Ἐρωδιός, 36, 37
Eul, xvii, 120, 121
Eul, Schleier, xvii, 130, 131 (*see* Ranseul)

Falco æsalon, xv
F. sacer, 208, 209
F. subbuteo, xvi
Falconidæ, xv, xvi
Falcula, 98–102
Fasant, xvii, 140, 141
Fasian, xvii, 140, 141
Fedoa, xvii, 44, 45
Feldefare, Feldfare, xviii, 58, 59
Fenclake, 194, 195
Fenlagge, 194, 195
Fenlake, 194, 195
Ficedula, 44, 45, 70, 71, 182, 183
Fieldfare, ix
Finch, 19, 119
Flasfinc, xvi, 159
Flasfinca, 158
Florus, 74, 75, 104, 105, 160, 161, 182, 183
Fowl, 201
Fox Goose, 25
Fringilla, 18, 52, 53, 72–75, 118, 130, 131, 148, 149, 182, 183
F. cœlebs, xvi
F. montifringilla, xvi
Fringillago, 130, 131
Fringillarius, 14, 15, 18, 19
Fringillidæ, xvi
Frugilega, 64, 65, 182, 183, 206, 207

Fulica, 70, 71, 74–79, 180, 181, 198
F. atra, 77
F. nigra xvii
Fulvia, 34, 35

Galerita, xiii, 18, 56, 57, 80–83
G. cristata, xiv
Galgulus (*Galgalus*), 88–91, 146, 147, 182, 183
Galli, 82, 83
Gallina, 86, 87, 138, 140, 148
G. cohortalis, 84
G. getula, 200
G. nigra, 32
G. nigra aquatica, 76
G. numidica, 84, 85
G. palustris, 76
G. rustica, 82, 84, 86
Gallinaceus, 180, 181
Gallinæ, 18, 82–87
G. africanæ (*africæ*), 82, 84, 86, 87
G. Hadrianæ, 84-87
G. villaticæ, 82, 84, 86
Gallinago, 40, 41, 86–88, 146, 147
Gallinula chloropus, xvii
Gallus, 82, 83
G. domesticus, xvii, 42
G. ferrugineus, xvii
G. gallinaceus, 200
G. medicus, 84
Γαμψώνυχες, 40
Ganss, xv, 22, 23
Ganss, Löffel, xvii, 150, 151
Ganss, Trap, xvi, 166, 167
Garrulus glandarius, xv
Gavia, 78, 110, 111, 180, 181
G. alba (White Gavia), 78, 180, 181
G. cinerea (Grey Gavia), 78, 180, 181
Gaviæ marinæ, 78
Gecinus viridis, xvii
Geelgorst, xvi, 106, 107
Geir, xviii, 176, 177
Γερανός, 94, 95
Gersthammer, xvi, 135
Gersthammera, 134
Geyr, xviii, 176, 177
Geyr Swalbe, xv, 102, 103
Ghiandaja, 145
Γλαύξ, 120, 121
Glede, xvi, 116, 117
Γλώττις, 104, 105
Godwit, ix
Godwitt, xvii, 45
Godwitta, 44
Goldfinc, 41
Goldfinca, 40
Goldfinche, xvi, 40, 41
Gold hendlin, xviii, 168, 169
Goose, 23, 25, 27, 29, 111, 123, 139, 195, 197

Goose, Bass, 31, 197
Goose, Bernacle (or Bernicle), x, 27, 31, 195
Goose, Brant (or Brent), xv, 27, 195
Goose, Fox, 25
Gose, xv, 22
Goshawk, 19
Gosling, 177
Gouke, xv, 66, 67
Graculi, 90, 91, 94, 95
Graculus, x, 90, 91, 94, 95
Graesmusch, xviii, 110
Grasmuklen, xviii, 44, 45
Grasmuschen, 47
Grasmuschus, 46, 72
Grassmusch (Grasmusch), xvii, xviii, 73, 136, 137
Grass-sparrow, 111
Greenfinch, 41
Grenefinc, 107
Grenefinca, 106
Grenefinche, xvi, 104, 105
Grey-clak, 195
Grey Hen, ix
Grey-Lag, 195
Grouse, Hazel, xvii
Gruidæ, xvi
Grünling, xii
Grunspecht, xvii, 89, 113, 115, 149
Grunspechta, 148
Grunspechtus, 88, 112, 114
Grus, 54, 62, 63, 94-96, 180, 181, 195
G. communis, xvi
Gryphen, 178
Gryps, 178, 179
Guinea Fowl, 85
Gull, 79
Gull, Black-headed, ix, 77
Gull, Grey, xvi
Gull, White, xvi
Γυπαίετος, 34
Γύψ, 176, 177
Guse, Solend, xvii, 28
Gypaëtus barbatus, xvi, 129

Hadrianæ, see *Gallinæ*
Hæmatopus, 102, 103
Halcyon, 20, 21
Haliæetos, 36, 37, 192, 193
Haliæetus, 34, 35
H. albicilla, xvi
Hän, xvii, 82, 83
Harrier, see Hen-Harrier and Marsh-Harrier
Hawk, 18, 19, 39, 68, 119, 167
Hedge-Sparrow, xvii, 136, 137
Hen, xvii, 82, 83
Hen, Getulian, 201
Hen-Harrier, ix, 19
Hen-Harroer, xv, 18
Hen, Marsh, 77
Hen, Mot, xvii, 170, 171
Hen, Water, xvii, 77, 170, 171
Heron, xv, 36, 37, 41, 55, 113, 125
Heron, Blue, 38
Heron, Cryel, xv, 38
Heron, Dwarf, xv, 38
Heron, Night, xiii
Hewhole, xvi, 115, 149
Hierofalco, 208, 209
Himantopus, 102
Hinnularia, 30, 31
Hirundinidæ, xvi
Hirundo, 96–102, 116, 117, 170, 173
H. agrestis, 98
H. (*alba*), 210
H. aquatica, xiii
H. domestica, 100
H. riparia, 98–102
H. rufula, 99
H. rustica, xvi, 98, 100, 101
Hobbia, 18
Hobby, ix, xvi, 19
Holtzsnepff, xvii, 86, 87
Holtztaube, xv, 60, 61
Hön, xvii, 82, 83
Horn Oul (Owl), xvii, 130, 131
Houp, xviii, 174, 175
Howlet, xvii, 120, 121
Howpe, xviii, 174, 175
Huhol, xvii, 89, 115
Huhola, 114, 148
Huholus, 88
Hydrochelidon nigra, xvi
Hyrundo, 204

Iaia, 144
Ibididæ, xvi
Ibis, Glossy, 57
Ibis, Red-cheeked, xvi
Icteros, 88, 89
Ἴκτινος, 116, 117
Iliacus, 170, 171
Iynx, 146–149
I. torquilla, xvii

Jay, xv, 145
Junco, 102, 103, 170, 171, 180, 181

Ka, xv, 92, 93
Kastrel, xvi, 166, 167
Kautz, xvii, 46, 47
Κέγχρις, 166, 167
Κελεός, 88, 89, 146
Κέπφος, 74-77
Kersenrife, xvi, 172, 173
Κίχλα, 170, 171
Κίγκλος, 54, 55
Kingfisher, xiv, 21, 23

Kirsfincke, xvi, 104, 105
Kistrel, xvi, 166, 167
Kite, x, 117, 193, 197
Kite (Blacker), 117
Κίττα, 142, 143
Κνιπολόγος, 64, 65
Koelmussh (Koelmusch), xvii, 136, 137
Koelmusshus, 136
Κόκκυξ, 66, 67
Kok of Inde, xvii, 86
Κολιός, 88, 89, 146, 147
Κολλυρίων, 58, 59
Kölmeyse, xvii, 130, 131
Κολοιός, 90, 91, 146
Κολυμβρίς (Κολυμβίς), 176, 177
Κορακίας, 90, 91
Κόραξ, 64, 65
Κορώνη, 64, 65
Κορυδαλός, 80, 81
Κόρυδος, 80–83
Κοττυφός (Κοπτυφός), 114, 115
Krae, xv, 64, 65
Kraeg, xv, 64, 65
Krammesvögel, xviii, 58, 59, 170, 171
Krän, xvi, 94, 95
Kränich, xvi, 94, 95
Κυανός, 52, 53, 181
Kukkuck, xv, 66, 67
Κύκνος, 82, 83, 120, 121
Kuningsgen, xviii, 152, 153
Kyngesfissher, xiv, 18, 19
Kyte, xvi, 116, 117
Kywit, xv, 77
Kywitta, 76

Læves, 14, 15
Lagopes (*Lagopus*), 104, 105
Λαγωφόνος, 34
Lagopus mutus, xviii
L. scoticus, xviii
Lämmergeier, 36, 129
Laniidæ, xvi
Lanius excubitor, xvi
L. minor?, xvi
Lapwing, x, xv, 175
Lapwinga, 76, 174
Laridæ, xvi
Lark, 19
Λάρος, 74, 75, 78, 79
Larus, xvi, 74, 75, 78, 79
L. ridibundus, xvi, 77
Laverock, xiv, 80, 81
Lefler, xvii, 150, 151, 153
Leflera, 152
Leporaria, 34, 35
Lerc, 81
Lerc, Wilde, 81
Lerca, 80
Lerch, xiv, 80, 81
Lerch, Heid, xiv, 80, 81
Lerck, Wod-, xiv, 80
Lerk, xiv, 80, 81
Lerk, Heth, xiv, 80
Lerk, Wilde, xiv, 80
Λευκερωδιός, 38, 39
Libicus, 164
Λιβυκός, 164, 165
Libyan Bird, 165
Ligurinus, 104, 105
L. chloris, xvi
Limosa belgica, xvii
Linaria, 158, 159
Lingett, xviii, 47, 111
Lingetta, 46, 110
Lingulaca, 104, 105, 128, 129
Linot, xvi, 51, 159
Linota, 50, 158
L. cannabina, xvi
Livia, 58–61
Löffel Ganss, xvii, 150, 151
Loun, xvii, 177
Louna, 176
Λύκος, 90, 91
Lüningk, xvi, 132, 133
Lupus, 90, 91, 94, 95
Luscinia, 44, 45, 68, 69, 108–111
Lutea (*Luteus*), 106–108
Luteola, 108, 109, 182, 183
Lyke Foule, xvii, 46, 47
Lyssklicker, xiii

Μαλακοκρανεύς, 116, 117, 132
Marcca penelope, xv, 22
Marsh-Harrier, 33
Marsh-Hen, 77
Martinette (or Martnette), Chirche, xv, 102, 103
Martinette, Rok, xv, 102, 103
Martnet, Bank, xvi, 102, 103
Matrix, 128, 129 (cp. 126)
Mavis, xviii, 172, 173
Medicæ, 84, 85
Meelmeyse, xvii, 130, 131
Melænaëtos, 34, 35
Μελαγκόρυφος, 44, 45
Μελανάετος, 34
Meleagris, 82–87, 140, 141, 200, 201
Melicæ, 84, 85
Mercolphus, xv, 144, 145
Mergus, xiv, 76, 77, 110, 111, 126, 127, 176, 177, 180, 181, 206, 207
Merl, xviii, 114, 115
Merlin, xv, 17
Merlina, 16
Meropidæ, xvi
Merops, 112–115
M. apiaster, xvi
Merula, 48, 49, 52, 53, 82, 83, 114, 115, 146–149, 164, 165, 170, 171
M. aquatica, xiii

Mewe, Wyss, xvi, 74, 75
Meyse, 130, 131
Meyspecht, xvii, 162, 163
Meyspechtus, 162
Miliaria, 50, 51, 158, 159
Milvius (*Milvus*), 16, 17, 116, 117, 192, 196
Milvus ictinus, xvi
M. ater (*nigrior*), xvi, 116
Mistletoe-Thrush, xviii
Molliceps, 58, 59, 116, 117, 120, 121, 132, 133, 168, 169
Monedula, 64, 76, 90-93, 146-149
Monetula, 90, 91
Monticola, 130, 131
Montifringilla, 72, 73, 182, 183
Moor-Buzzard, ix, 33
Moorhen, 171
Morhen, xviii, 87, 171
Morhenna, 86
Morinellus, 202-205
Morphna (*Morphnos*), 30, 31
Mortetter, xviii, 159
Mortettera, 158
Motacilla, 64, 65
M. alba, xiii, xvi
M. lugubris, xvi
Motacillidæ, xvi
Mot Hen, xvii, 170, 171
Müsche, xvi, 132, 133
Myre Dromble, xv, 38

Nachtgäl, xvii, 108, 109
Nachtra, xii, xiii
Nachtrab, xiii
Nachtrap, xii
Naghtrauen, xii, xiii
Νεβροφόνος, 30
Νῆττα, 22, 23
Νηττοφόνος, 30
Neun mürder, xvi, 168, 169
Neunmürder, Grosse, 168, 169
Night-Heron, xiii
Noctua, 46, 47, 120, 121, 130, 131, 180, 181
Nonna, 132
Nousbrecher, xv, 94
Nucifraga, 94, 95
N. caryocatactes, xv
Nucipeta, 162
Nuin mürder, xvi, 116, 117
Νυκτικόραξ, 120, 130
Numida meleagris, xvii
Numidica, see *Gallina*
Nun (Non), xvi, 133
Nushäkker, xvii, 162, 163
Nutcracker, x
Nutjobber, xvii, 162, 163
Nutseeker, 163
Nycticorax, xii, xiii
Nyghtyngall, xvii, 108, 109
Nyn murder, xvi, 116, 117, 168, 169
Nyroca ferina, xv

Οἰνάς, 62, 63
Oistris, xvii, 164, 165
Olor, 120-123, 126, 180-183
Onocrotalus (*Onocratalus*), x, xiii, xiv, 122-127
Orfraie, 193
Oriolidæ, xvi
Oriolus galbula, xvi
Ὀροσπίζης, 72, 73
Ὄρτυξ, 62, 63
Ortygometra, 126-129
Osel, Black, xviii, 114, 115
Osprey, ix, xvi, 34, 35, 37, 193-195
Ossifraga, 36, 118, 128, 129, 193
O. barbata, 128
Ossifrage, 37, 129
Ossifrage, Bearded, 129
Otididæ, xvi
Otis, 130, 131, 168, 169
O. tarda, xvi
Ὠτὸς, 120, 130, 131
Otus, 128-131, 204, 205
Ὄυραξ, 166, 167
Ousel, Water, 23
Owl (Oul), xvii, 120, 121
Owl, Eagle, 47
Owl, Hawk, 157
Owl, Horn, xvii, 130, 131
Owl, Horned, 120
Oxei, Great, xvii, 130
Oxeye, Great, 131

Pagelün, xvii, 136, 137
Palmipedes, 120
Palumbarius, 14, 15, 18, 19, 56, 57
Palumbes, 58-61, 68, 69, 180, 181
Palumbus, 18, 60, 61
Pandion haliaëtus, xvi
Pandionidæ, xvi
Papegay, xvii, 150, 151
Paphus, 50, 51
Πάρδαλος, 58
Pardalus, 132, 133
Paridæ, xvi
Parrot, xvii, 209
Partridge, 43, 63, 141, 203
Parus, 130-133, 182, 183
P. ater, xvii
P. cæruleus, xvi
P. major, xvii
P. maximus, 132, 162
P. medius, 130
P. minimus, 132
P. palustris, xvii
Passer (*Pascer*), 18, 72, 104, 106, 108, 137-139, 160, 168, 180-183

Passer communis, 134
P. domesticus, xvi
P. gramineus, 110, 111
P. harundinaceus, 104, 105
P. magnus, 134
P. sepiarius, 136, 137
P. torquatus, 134, 135
P. troglodites (*troglodytes*), 134–137
Passer, Common, 135
Passer, Great, 135
Passerculus, 134
Pavo, 136–139
P. cristatus, xvii
P. indicus, 86
Peacock, 137, 139
Peacock, Indian, 87
Pecok, xvii, 136, 137
Πελειάς, 60, 61
Pelecanus (*Pelicanus*), 150–153
Πελεκάν, 150, 151
Pella, 38, 39
Penelopes, 22–25
Penelops, 22, 182, 183
Percæ, 14, 15
Percnopterus, 34, 35
Percnos, 30–33
Perdix, xiv, 18, 42–44, 138–141, 180, 181, 202
P. cinerea, xvii
P. rustica (*rusticula*), xii, 140, 141
P. vulgaris, xiv
Πέρδιξ, 138, 139
Περιστερά, 58, 59
Περιστεροειδῆ, 40
Πέρνης, 14
Pernix, 14, 15
Pertrige, xvii, 138, 139
Pffaw, xvii, 136, 137
Phalacrocoracidæ, xvii
Phalacrocorax, x, 92–95
P. carbo, xvii
P. graculus? xvii
Phalaris, 92, 93, 182, 183
Phasianidæ, xvii
Phasianus, 40, 140, 141, 200
P. colchicus, xvii
Φάττα, 60, 61
Pheasant, 41, 201
Φήνη, 36
Phesan, xvii, 140, 141
Philomela, 108, 109
Phœnicura, 157
Phœnicurus, 154–157
Phœnix, xvii, 140, 141
Φοινικουρός (φοινικούργος), 154, 155
Phoix, 38–41
Φῶϋξ, 38
Pica, 118, 142–145, 170, 171
P. granata, 144
P. marina, 198
Pica rustica, xv
P. vulgaris, 144
Picidæ, xvii
Picus, 54, 114, 115, 146–149, 162, 163
P. cornicinus, 148
Picus, Crow, 149
Picus, Green, 115
Picus martius, xiii, xvii, 88, 89, 146, 147, 182, 183
P. maximus, 148
P. medius, 148
P. minimus, 148
P. viridis, 112
Pie, 119, 145, 199
Pie, Common, 145
Pie, Sea, 199
Pie, Seed, 145
Pigargus (*Pygargus*), xii, 30, 31
Pigeon, 43, 45, 195
Piger, 38, 39
Pilaris (*Turdus*), 170, 171
Piot, xv, 142, 143
Pipers (of Crane or Pigeon), 96, 97
Pipiones, 96, 97
Pipit, Tree-, 69
Pipo, 146, 147
Pittour, xv, 40, 41, 123
Pittourus, 40, 122
Πλάγγος, 30
Plancus, 30, 31
Planga, 30, 31
Πλάνος, 30
Platalea leucorodia, xvii, 41
Plataleidæ, xvii
Platea, 150, 151, 180, 181
Platelea, 150, 151
Plegadis falcinellus, 57
Plover, Stilt, 102
Pluver, xv, 132, 133
Pochard, xv, 49
Pochardæ, 48
Podicipedidæ, xvii
Podicipes minor, xvii
Popinjay, xvii, 150, 151
Porphyrio, 102, 103, 152, 153
P. cœlestis, xvii
Pratincola rubicola, xviii
Πρέσβυς, 152, 153
"Priest," 51
Procellariidæ, xvii
Ψάρος, 164, 165
Psitaca, 150
Psitace, 150, 151
Psitacus (*Psiticus*), 82, 83, 142, 143, 150, 151
Psittacidæ, xvii
Psittacus (*Sacropsittacus*), 208–211
Πτέρνης, 14
Πτερνίς, 14
Ptynx, 56, 57

Puffinus, xvii
Πύγαργος, 170; see *Pigargus*
Pulla, 34, 35
Pulver, xv, 132, 133
Puphin, 205
Puphinus, 204
Pupin, 197, 205
Pupinus, 196, 204
Πυῤῥούλας, 160, 161
Πυῤῥουράς, 161
Puttok, xvi, 116, 117
Py, xv, 142, 143
Pygargus, see *Pigargus*
Pygosceles, 204, 205
Pyrrhocorax, 90–93
P. alpinus, xv
P. graculus, xv
Pyrrhula europæa, xvi

Quail, 63, 69, 203, 205
Quale, xviii, 62, 63
Querquedula crecca, xv
Quikstertz, xvi, 64, 65

Rabe, xv, 64, 65
Rail, Water, 71
Rala, xii, xiii, 140
R. aquatica, xii
R. aquatilis, xiii
R. montana, xiii
R. sylvestris, xiii
R. terrestris, xii
Rale, xvii, 141
Rallidæ, xvii
Rallus aquaticus, 71
Ranseul, xvii, 130, 131
Raphön, xvii, 138, 139
Rauce, 206, 207
Raucus, 206, 207
Raven, xv, 35, 64, 65
Raven, White, 211
Rayn byrde, xvii, 146
Redbreast, ix, 161
Redbreste, Robin, xvii, 154, 155
Rede Sparrow, xvi, 102, 103
Rede tale, xviii, 154, 155
Redshanc, xvii
Redshanca, 102
Redshank, 103
Redstart, ix, 155
Reed-Bunting, x
Reed-Sparrow, 105
Regulus, 52, 53, 118, 119, 134–137, 152–155, 168, 169, 182, 183
R. cristatus? xviii
Reydt Müss, xvi, 102, 103
Reyger, xv, 36, 37
Ringel Taube, xv, 60, 61
Ringged Dove, xv, 60, 61
Ringtail, 19
Ringtale, xv
Ringtalus, xii, 18
Ringtayle, xii
Riparia, 98–102
Robin Redbreste, xvii, 154, 155
Rook, 65
Rosdom, xv, 40, 41, 123
Rosdomma, 122
Rosdommus, 40
Rötbrust, xvii, 154, 155
Rötkelchen, xvii, 154, 155
Rötstertz, xviii, 154, 155
Rouce (Rouke), 206, 207
Rowert, xvi, 72, 73
Rubecula, 154–159, 182, 183
Rubetarius, 14, 15, 18, 19
Rubetra, 158, 159, 182, 183
Rubicilla, 156, 157, 160, 161, 182, 183
Rupex, 180, 181
Ῥυσομήτρης, 40, 41
Rusticula, xii
Ruticilla, 154–159
R. phœnicurus, xviii

Sacropsittacus, 208–211
Salus, 160, 161
Sandpiper, ix
Sandpiper, Common, 57
Saxicola œnanthe, xviii
Σχιζόποδα, 40
Σχοίνικλος, 102, 103
Schrecke, 71
Schric, xvi, 169
Schricus, 168
Schriek, 71
Schryk, xvii, 70, 71
Schuffauss, xvii, 46, 47
Schüffel, xvii, 46, 47
Schwalb, xvi, 96, 97, 102, 103
Scolopacidæ, xvii
Scolopax rusticula, xvii
Scops, 204, 205
Scrica, xvii, 128, 129
Sea-Aquila, 183
Sea-Cob, 79
Sea-Eagle, 193
Sea-Pie, 199
Se Cob, xvi, 78–79
See Gell, xvi, 78–79
Σεισοπύγις, 54, 55
Semaw (white with a black cop), xvi, 74, 75
Senator, 152, 153
Serinus canarius, xvi
Shearwater, 71
Sheldappel, xvi, 72, 73
Sheld-Drake, 25
Shell-apple, 73
Shovelard, xvii, 41, 150, 151
Shovelarda, 38

Shric, 169
Shricus, 168
Shrike, ix, xvi, 116, 117
Siskin, xvi, 51, 108, 109
Sitta, 162, 163
S. cæsia, xvii
Sittidæ, xvii
Σκνιποφάγα, 40
Σκολόπαξ, 87
Σκωληκοφάγα, 40
Smatche, xviii, 52, 53
Smerl, xv, 17
Smerla, 16
Snepff, Holtz, xvii, 42, 86
Sneppa, 70, 71
Solend Guse, xvii, 28
Sparhauc, xv
Sparhauca, 18
Sparrow, xvi, 73, 105, 107, 109, 111, 132, 133, 135, 161, 169
Sparrow, Grass, 111
Sparrow-Hawk, 19
Sparrow, Hedge, xvii, 136, 137
Sparrow, Rede, xvi, 102, 103
Sparrow, Reed, 105
Spatz, xvi, 132, 133
Specht, xvii, 146, 147, 149
Spechta, 148
Speiren, xvi, 102, 103
Sperlingk, xvi, 132, 133
Σπερμολόγος, 64, 65, 206, 207
Spermologus, 206, 207
Sperwer, xv, 18, 19
Spink, xvi, 72, 73
Spinus, 40, 41, 64, 65, 104–107, 182, 183
Σπίζα, 72, 73
Spoonbill, x, 41
Stär, xvii, 164, 165
Starling, 23, 95, 165, 205
Στεγανόποδα, 40
Steinbeisser, xiii
Steinchek, xviii, 52, 53
Steingall, xvi, 166, 167
Stellaris, 14, 15, (*Ardea*) x, 38-41, 126, 127
Sterlyng, xvii, 164, 165
Stern, xvi, 79
Sterna, 78
S. nigra, 79
Steynbisser, xvii, 54, 55
Stigelitz, xvi, 40, 41
Stilt-Plover, 102
Stocdove, xv, 60, 61
Stonchatter, xviii, 159
Stonchattera, 158
Stör, xvii, 164, 165
Storck, xv, 54, 55
Stork, xv, 54, 55
Strauss, xvii, 164, 165
Strigidæ, xvii
Strix stridula? xvii
Στρούθος, 132, 133
Στρουθός, 164, 165
Struthio, 44, 45, 164, 165
Struthio camelus, xvii, 92, 93, 164, 165, 168, 169
Struthionidæ, xvii
Sturnidæ, xvii
Sturnus, 22, 94, 164, 165, 168, 169, 204
S. vulgaris, xvii
Subaquila, 34, 35
Subuteo (*Subbuteo*), xii, 14, 15, 18, 19
Συκαλίς, 70, 71
Sula bassana, xvii
Sulidæ, xvii
Swalbe, Geyr, xv, 102, 103
Swalbe, Kirch, xv, 102, 103
Swalbe, Über, xvi, 102, 103
Swale, xvi, 96, 97
Swallow (Swallowe), ix, xvi, 96, 97, 101-103, 205
Swallow, Black, 101
Swallow, Great, xv, 102, 103
Swallow, House, 101
Swallow, Water, xiii, xvii, 54, 55
Swan, xv, 120, 121, 123, 127, 197
Swän, xv, 120, 121
Sylvia, 182, 183
S. atricapilla, xviii
S. rufa, xviii
Sylviidæ, xvii, xviii

Tadorna cornuta, xv, 25
Ταών, 136, 137
Tarda, xii, 180, 181, (see also *Aves tardæ*)
Taube, xv, 58, 59
Taube, Ringel, xv, 60
Tela, 48
Tele, xv, 49
Tern, Black, ix, 79
Tetrao, 30, 31, 130, 131, 166, 167
T. tetrix, xviii
Tetraonidæ, xviii
Tetrix, 30, 31, 166, 167
Τέτριξ, 166, 167
Thraupis, xii, 52, 53
Θραυπίς, xii, 50, 51
Throssel, xviii, 172, 173
Thrusche, xviii, 170, 171
Thrush, 59, 119, 175, 177, 209
Thrush, Blue, 53
Thrush, Mistletoe, xviii
Thrushe, xviii, 172, 173
Tichodroma muraria, 53
Tinunculus (*Tinnunculus*), 166, 167
Tinnunculus alaudarius, xvi
Titlark, 69

Titling, 69
Titlinga, 68
Titlyng, xvi, 68, 69
Titmouse, 130, 131
Titmous[e], Great, xvii, 130, 131
Titmouse, Greatest, 133
Titmous[e], Less, xvii, 130, 131
Torquella (*Torquilla*), 146–149
Totanus calidris, xvii
Trapp (Träp), xvi, 31, 166, 167
Trappus, xii, 30
Tree-Creeper, ix
Tree-Pipit, 69
Triorcha, 14, 15
Triorches, 16, 17
Τριόρχης, 14, 16, 17
Τροχίλος, 152, 153
Trochilus, 152, 153
Troglodytes (*Troglodites*), 134–137
T. parvulus, xviii, 155
Troglodytidæ, xviii
Τρύγγας, 170, 171
Τρυγών, 60
Trynga, 170, 171, 180, 181
Turbo, 146–149
Turdidæ, xviii
Turdus, 58, 102, 105, 118, 170–174, 208
T. iliacus, xviii, 170, 171
T. merula, xviii
T. musicus, xviii
T. pilaris, xviii, 170, 171
T. viscivorus, xviii, 170, 171
Turtel Duve, xv, 60, 61
Turtle-Dove, 89
Turtur, 58–61, 88, 89, 146, 147, 172–175, 180, 181
T. communis, xv
Tyrannus, 120, 121, 154, 155, 168, 169

Üle, xvii, 120, 121
Ulula, 130, 131
Ὑπαιετός, 34
Upupa, 174, 175
U. epops, xviii
Upupidæ, xviii
Urax, 166, 167
Urinatrix, 92, 93, 176, 177, 182, 183

Valeria, 34, 35
Vannellus, 76, 77, 174, 175
V. vulgaris, xv
Velt hön, xvii, 138, 139
Venice Dove, xv, 63
Villaticæ, 82, 84, 86
Vinago, 58–63, 180, 181
Vipiones, 96, 97
Vireo, 58, 59, 106, 107, 148, 149, 172–175
Viscivorus (*Turdus*), 170, 171
Vishärn, xvi, 34, 35
Vogelhain, xiii
Vulpanser, xii, 24, 25, 182, 183
Vultur, xviii, 34, 166, 176–178
V. niger, 178
Vulture, 35, 177, 179
Vulture, Black, 179
Vulturidæ, xviii

Wachholtervögel, xviii, 170, 171
Wachtel, xviii, 62, 63
Wagtale, xviii, 64, 65
Waldrapp, x
Wall-Creeper, 53
Waltrap, xvi
Walt-rapp, x
Waltrapus, 92–95
Wasser Hen, xvii, 170, 171
Wasser Steltz, xvi, 64, 65
Water-Craw, xv, 22, 23
Water-Hen, xvii, 170, 171
Water-Hen (black with a white frontal patch), 77
Water Ousel, ix, 23
Water Rail, 71
Water Swallow, xiii, xvii, 54, 55
Weidwail, xvi, 172, 173
Weingaerdsvogel, xviii, 172, 173
Weye, xvi, 116, 117
Wheatear, ix, 53
Widhopff, xviii, 174, 175
Wigene, xv, 48, 49
Wigeon, 22
Witwol, xvi, 172, 173
Wodcoccus, 42
Wodcok, xvii, 86, 87
Wodlerck, xiv, 80
Wodspechta, 148
Woodcock, 42, 45, 89
Woodpecker, 55
Woodspecht (Wodspecht), xvii, 149
Wren, xviii, 152, 153, 155
Wyngthrushe, xviii, 172, 173

Yelowham, xvi, 106, 107
Yowlring, xvi, 106, 107

Zaunküningk, xviii, 152, 153
Zeysich, xvi, 108, 109

www.ingramcontent.com/pod-product-compliance
Ingram Content Group UK Ltd.
Pitfield, Milton Keynes, MK11 3LW, UK
UKHW040705180125
453697UK00010B/412